Hegemony, Security Infrastructures and the Politics of Crime

This book examines the politics of crime and the response to it in Potchefstroom, a small settler colonial city in South Africa. It draws on the city's everyday practices and experiences to offer local bottom-up insights into security beyond the state.

The book provides a comprehensive understanding of security beyond the state and how security workers and residents experience and perceive their own security practices, their daily interactions with other security providers which influences power dynamics between those who express fear through various platforms and those deemed potential criminals. It aids in re-conceptualising violence and security governance in South Africa with a view to analysing the processes of crime prevention and management, the changing nature of public and private spaces and how these spaces interact with state and local authorities. In a rigorous exploration of the ways to tackle the complex problem of crime, the book critiques an overreliance on security infrastructures such as social media, gated barriers, neighbourhood residents' associations and private security companies. It also looks at how crime is treated as an individual as opposed to a societal problem. The book addresses the urgent need for collaboration across these fault lines to promote a more inclusive security in a broader fragmented social and political context.

With a novel analytical approach based on the twin optics of infrastructure and post-structural hegemony, the book will be relevant to scholars and students of South African politics and critical security studies, as well as the international audience interested in crime and private security.

Gideon van Riet is a senior lecturer in political studies at the North-West University in South Africa. His research focusses on security as it pertains to disasters and crime. Gideon's first monograph, *The institutionalisation of disaster risk reduction: South Africa and neoliberal governmentality*, was published by Routledge in 2017.

Routledge Studies in Urbanism and the City

For more information about this series, please visit:
www.routledge.com/Routledge-Studies-in-Urbanism-and-the-City/book-series/RSUC

Hegemony, Security Infrastructures and the Politics of Crime

Everyday Experiences in South Africa

Gideon van Riet

LONDON AND NEW YORK

First published 2022
by Routledge
2 Park Square, Milton Park, Abingdon, Oxon OX14 4RN

and by Routledge
605 Third Avenue, New York, NY 10158

Routledge is an imprint of the Taylor & Francis Group, an informa business

British Library Cataloguing-in-Publication Data
A catalogue record for this book is available from the British Library

Library of Congress Cataloging-in-Publication Data
A catalog record for this book has been requested

ISBN: 978-0-367-46326-7 (hbk)
ISBN: 978-1-032-12003-4 (pbk)
ISBN: 978-1-003-02818-5 (ebk)

DOI: 10.4324/9781003028185

Typeset in Times New Roman
by Apex CoVantage, LLC

Contents

Illustrations

Figures

Maps

Tables

Acknowledgements

I express my gratitude to the following individuals and organisations who kindly gave me their time and support. These include:

- Johann and Christiaan Hafeale and the staff at Mooirivier Beskerming;
- Oskin and Garry Chatwind and the staff at CSS Security;
- Botes Botha and Theuns Jacobs at CB Security;
- The members of Sector One and Sector Two of the Potchefstroom community policing forum;
- Mr Des Ayob from North-West University Protection Services and the community policing forum's Sector Four;
- All other interviewees and participants otherwise, not named to protect their identities and those whom I may have omitted by accident (my sincere apologies).

I thank my colleagues who served as sounding boards for my ideas during formal and informal engagements over many years, or who generously gave their time to read earlier drafts of parts of this work. I have over the years benefitted from discussions with Pieter Heydenrych, Andre Goodrich, Pia Bombardella, Cecilia Schultz and Andreas Langenohl, while inputs more specific to this book came from Anné Verhoef, Willem Mostert, Johan Zaaiman, Don Wallace, Norman Sempijja, Peter Vale, Rita Abrahamsen and two anonymous peer reviewers. All interpretations and mistakes are of course my own.

I also acknowledge the Social Sciences Research Council (SSRC)/African Peacebuilding Network (APN) for funding my research in the neighbouring Ikageng police precinct through a grant for the project entitled, 'Transitional injustice: The spatial politics of crime in contemporary South Africa' – 2019. The data from that project have been immensely valuable in interpreting this book's subject matter.

Finally, I thank my wife Cecilia Schultz for her support, even while she was in the midst of her own PhD studies.

Abbreviations

ANC	African National Congress
ASGISA	Accelerated and Shared Growth Initiative for South Africa
ATKV	*Afrikaanse Taal en Kultuur Vereniging* (Afrikaans Language and Culture Association)
AWB	*Afrikaner Weerstandsbeweging* (Afrikaner Resistance Movement)
BBBEE	Broad-based black economic empowerment
BEE	Black economic empowerment
CB	CB Security
CBD	Central business district
CCPS	Cape Town Central Police Station
CCTV	Closed-circuit television
CID	City improvement district
COP	Community-oriented policing
CP	Conservative Party
CPF	Community policing forum
CPS	Campus Protection Services (of the North-West University)
CPUT	Cape Peninsula University of Technology
CSS	Critical security studies
EBP	Evidence-based policing
EFF	Economic Freedom Fighters
FDI	Foreign direct investment
GEAR	Growth Employment and Redistribution
GIS	Geographical information systems
IDP	Integrated development plan
IPS	International Political Sociology
IPSA	International Political Studies Association
ISI	Import substitution industrialisation
ISP	Internet service provider
JLU	Justus Liebig University
JSE	Johannesburg Stock Exchange
MB	*Mooirivier Beskerming* (Mooi River Protection)
MMA	Mixed Martial Arts
NCCS	National Crime Combatting Strategy

NCPS	National Crime Prevention Strategy
NDP	National Development Plan
NGP	New Growth Path
NIMBY	Not in my backyard
NP	National Party
NWU	North-West University
PARIS	Political anthropological research for international Sociology
PSC	Private security company
PSIRA	Private Security Industry Regulating Authority
PU for CHE	Potchefstroom University for Christian Higher Education
PUK	*Potchefstroomse Universiteitskollege* (Potchefstroom University College)
RDP	Reconstruction and Development Programme
SANDF	South African National Defence Force
SAP	South African Police
SAPS	South African Police Service
SDF	Spatial development framework
SOE	State-owned enterprise
SPSS	Statistical Package for Social Scientists
SRWP	Socialist Revolutionary Workers Party of South Africa
SSRC/APN	Social Sciences Research Council/ African Peacebuilding Network
SU	Stellenbosch University
SUV	Sports utility vehicle
TRT	Tactical Response Team of the SAPS
UCT	University of Cape Town
UFS	University of the Free State
UP	University of Pretoria
ZAR	*Zuid-Afrikaansche Republiek* (South African Republic)

Preface

In 2016 I attended the International Political Studies Association's (IPSA) conference in Poznan, Poland. I was about to submit to the publisher the final version of a book manuscript based on my doctoral thesis. One night, Laurence Piper, a senior South African political scientist, and I met for drinks. As I had effectively finished publishing my doctoral research on disaster risk reduction in South Africa, I was looking for an additional long-term research interest. There were a few contenders. Laurence encouraged me to pursue one of these emerging interests, namely the role of private security companies (PSCs), in particular armed response companies, in everyday South African politics. The literature on PSCs had, as I soon realised, become quite saturated. However, the more I researched, the more I realised that these companies were merely one part of a larger whole. The question I had was how could I frame such a holistic approach to the politics of crime. Some answers were to be found in Criminology, but these could not be adopted without reflection by a political scientist/political sociologist such as myself: someone who foregrounds societal power relations.

Here the constructive influence of Andreas Langenohl of Justus Liebig University (JLU) in Germany, and his colleagues in the Collaborative Research Centre, Dynamics of Security, was immense. We met in 2013 because of a cooperation agreement between our respective universities. Since then we have frequently been collaborating formally and informally. Through this collaboration I became acquainted with the two theoretical pillars of this book: infrastructure as a topic for social science research and the work of Ernesto Laclau (initially with Chantal Mouffe) on hegemony. I recognised the potential in combining an infrastructural imaginary with a post-structural interpretation of hegemony in analysing the politics of crime and the response thereto. This approach offers insight into the flow of security practices through diverse physical and ontological structures. Moreover, such a perspective potentially offers a path for critical scholarship beyond mere deconstruction, towards productive reflection on praxes and in aid of radical democracy, founded in a realisation of the radical contingency of the social. Radical contingency encourages non-dichotomous conceptions of social antagonism based on an understanding of the non-fixity and non-uniformity of individual, group and organisational identities. This lack of uniformity is where I believe *hope* for a better social order is to be found. So, this book focussed on crime, but deeply concerned with South African politics and a (set of) progressive post-apartheid South African project(s), does make tentative suggestions that whilst pertinent to the politics of crime hopefully extend more broadly. The discussion up to this point accounts for how and why this book came about. However, there were various events in 2020 that necessitated this Preface, which likely would not even have existed otherwise.

While writing this book, some events unfolded that will likely have a major bearing on the discipline of critical security studies (CSS) from which I am writing, and more generally, the organisation of societies at various scales. I would be remiss if I did not offer this Preface as an explanation of how the book should be viewed in relation to these developments. Let me be clear, most of these developments are not new; they simply gathered momentum in 2020. The matters in question are the Covid-19 pandemic, the murder of George Floyd in Minneapolis and a similar murder of Collins Khoza in South Africa. Finally, there was the 'securitisation debate' within CSS. This debate ignited after the publication of the second of two critiques of (Western) critical theories and their application within CSS, for being racist.

While the implications of the Covid-19 pandemic are still quite uncertain, I can speak more clearly on how the latter two and their consequences bring into sharp relief the strengths and weaknesses of this book. I consider these two debates important discussions to which I can only offer provisional positions. Aspects of what these events represent and what the fallout might be are incorporated, to a degree, into the book I was already in the process of writing at the time. These attempts may not be enough for some and I welcome critical engagement based on such potential inadequacies. While the latter two events have sparked vigorous debates within the media, academia and the Twittersphere, this book deals with a particular Southern case study, which to a certain degree complicates some of the arguments made in the North, at least in their most extreme forms. However Southern as this case study might be, it is still written by a white South African man, in fact, an Afrikaans first language, white South African man. This is a historically notorious category, although I would contest that as with all other categories, it is not homogenous. Still, this book on crime and therefore, as it will be argued, largely on inequality has been written by someone in a position of relative privilege in one of the most materially unequal countries on the planet. South Africa is also a country that despite mythical notions of a 'miracle' transition is to this day characterised by incessant white on black racism and racist structures and circulating discourses that have been reiterated and enforced through everyday practices. Perpetrators now also include people of colour, especially those in positions of relative power or those who have through their employment become cogs in the wheel of racist (re)articulations. Therefore, to summarise, the notion of positionality, in this case pertaining to relative security and freedom from excessive policing, is not lost on me.

This book does foreground racial and other inequities within South Africa, be it by drawing mostly on Western theorists. I do maintain, provisionally, that one can use such theorists for anti-racist purposes. They may also inform productive alliance-building across historical and reiterated fault lines. Furthermore, perhaps we should be reconsidering critique as an end in itself and ask, as James Ferguson (2009) does, 'what do we want?'. What type of social order(s) are we after, bearing in mind that these might be moving targets? How will we get there or instil the types of socialisation and material changes we want? In the process critique will seemingly be an essential building block for more meaningful and academically more courageous conclusions.

With all of these said, scholarship ought to be an ethical enterprise. Ethics in terms of fairness to all participants and implicated communities has been an objective of this work. This ethics also extends to the people being studied. In this case it is mostly the wealthier segments in society and how they respond to crime. I have attempted to be empathetic or at least sympathetic to all of these groups and organisations, even where I fundamentally disagree with them. Practically, this has meant trying to understand the reasoning they employ and to acknowledge positive aspects of their work. Such an approach is possibly more likely to lead to productive deliberation across prevalent fault lines. In contrast, dismissive analyses would be both unethical and unproductive. Moreover, it is arrogant to assume superior, final knowledge.

The murders of George Floyd and Collins Khoza have once again brought to the fore the abolitionist discourse on criminalisation and policing. This perspective holds that policing and criminalisation is but a fraction of any meaningful strategy for crime reduction and security provision. As such, abolitionism, especially when not taken entirely literally in the short term, is quite compatible with the argument I was already committed to. The main difference is that this book is written in a context different from that of authors such as Alex Vitale (2017). This book is a critique, which makes some general suggestions about 'the good social order' that we should be striving for. Following Laclau this would be a social order that is likely to emerge in increments, stemming from multiple and often overlapping progressive praxes. What an abolition democracy as per W.E.B. du Bois ([1935] 2017) and Angela Davis (2005) might look like in a context such as South Africa is an intriguing and seemingly important avenue for future research. However, the financial fallout of the Covid-19 pandemic, judging on recent assessments of the South African economy, will require some very creative thinking. Such a brand of deep engagement falls outside of this book's scope. This book might, however, contribute towards a foundation for further analyses along abolitionist lines.

Gideon van Riet
Potchefstroom
March 2021

References

Davis, A.Y. 2005. *Abolition democracy. Beyond empire, prisons and torture. Interviews with Angela Davis*. New York: Seven Stories Press.

Du Bois, W.E.B. 2017[1935]. *Black reconstruction in America*. London: Routledge.

Ferguson, J. 2009. The uses of neoliberalism. *Antipode: Journal of Radical Geography*, 41(s1), pp. 166–184.

Vitale, A.S. 2017. *The end of policing*. London: Verso.

Introduction

Crime, security and politics

This book reflects certain conjunctures between the fields of critical security studies (CSS) and studies of crime. The work is located between an extended view of security and an extended view of policing, or the study of policing as opposed to purely a study of the police. The project started as a general investigation into private security companies (PSCs), in particular those engaged in armed response, as key intermediaries in the politics of contemporary South Africa. Much of this focus remains, but PSCs have been moved slightly from the centre of the investigation to account more fully for the sociality of crime and the response thereto. Private security does not operate in isolation. These organisations are intricately intertwined with the realities of contemporary South Africa. The problem of crime and crime fighting operates with and through various platforms and organisations, such as community policing structures, social media, the print media and environmental design. These practices reveal the immense divisiveness of the politics of crime.

The book takes inspiration from International Political Sociology's (IPS's) assertion that security is not simply a matter of 'high' politics. It manifests in daily life in many ways. Security, contra the Copenhagen School (cf. Buzan et al., 1998), is not simply about existential threats, although some actors discussed in the book often view crime in this way, especially as it relates to themselves. Crime as a security issue is often portrayed as threatening particular ways of life and as state decline. Consequently, some residents seek to insulate themselves from crime with various degrees of success. In Addition, the book seeks to answer what Bigo refers to as the main questions of IPS, namely what does security mean and what does it do? (Bigo, 2008:116). The book does this by focussing on a field less often studied in Security Studies. Unlike the likes of Bigo, it does not draw on Field Theory. The International Relations literature that deals with PSCs tends to draw on assemblage theory or Bourdieusian modes of analysis to make sense of the complexity of security provision (cf. Abrahamsen and Williams, 2009; Gheciu, 2015). These authors tend to emphasise how security provision mirrors global (neoliberal) transformations over the past 50 years. Because there is an artificial divide between the international and the national security in mainstream Security Studies, the Sociology of Crime and Deviance, and Criminology is regularly neglected by this field (Bigo, 2008:118).

DOI: 10.4324/9781003028185-1

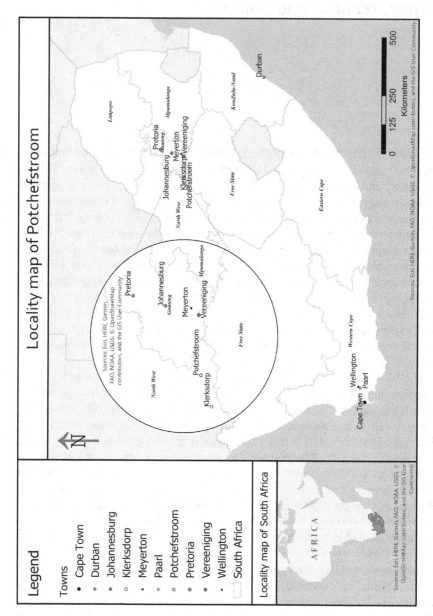

Map 1.1 Location map of Potchefstroom

In addition to CSS, the book is also located within the debates on PSCs and urban forms associated with the aforementioned insularity. The criminological literature seems to be divided roughly into two camps (Diphoorn, 2015a:198). The first posits nodal governance as a descriptor for how crime is dealt with (cf. Shearing and Wood, 2003). This implies a relatively 'flat ontology', where authority is less hierarchical and the state police becomes one amongst many *relatively* equal actors in crime fighting. These authors often focus on one or two nodes only in their analyses, whereas others disagree and draw on network governance as a means of understanding crime fighting (cf. Du Pont, 2004). These authors emphasise the relationship and interactions between nodes. In the case of South Africa, the notion of *semigration* (Ballard, 2004) would resonate with either of these two approaches. High walls, alarms and security estates are meant to insulate residents from the realities on the other side of these infrastructures. Some of those who posit nodal governance also talk of 'mass private property' (Shearing and Stenning, 1981), whereby the privatisation of public spaces has influenced how policing takes place. Yet authors such as Jones and Newburn (1999) indicate that the influence and degree of novelty in new property regimes on the rise of private security is unproven.

In the case of South Africa, the rise of private security can be traced to the latter decades of apartheid and the engagement of the police in stemming the insurgent politics that led to successive states of emergency. Security companies became a resource for protecting existing forms of property ownership. With democracy and the emergence of numerous shopping malls around the country, private security did, however, become a means of protecting newer property arrangements, and there is a significant political economy of crime and property in post-apartheid South Africa that seemingly resonates well with the notion of mass private property. Although this is not the focus of this book per se, it should be stated for the sake of contextualisation that since the end of apartheid, fears of integration between different segments of society and rising crime have fuelled greater insularity, and more gated communities have been developed. There is vast literature on gated communities (cf. Atkinson and Blandy, 2005; Low, 2001). Similarly, there has been an exponential growth in shopping malls on private land with their own security systems. Listed companies that develop shopping centres became rather lucrative investments at one point, although saturation may have more recently reduced returns on shareholder's investment.

The literature on private security has developed significantly in recent years. Much of the literature pertains to private military companies and the so-called market for force (cf. Percy, 2007; Avant, 2005). The current project is less concerned with that body of literature. For our immediate purposes what is of interest is the literature on armed response companies. In this matter South Africa is a significant, even archetypical, case study. Diphoorn's (2015b) ethnographic work in Durban is arguably one of the, if not the most, comprehensive work on the topic. Backed up by extremely thorough fieldwork, she gives much insight into this sector. Themes covered by her include, but are not limited to, the bodily capital of private security operatives and their ability to work from positions of authority, the racialised nature of the industry in terms of internal racism, the racial stereotyping of suspects and the problematic nature of notions of sovereignty. One important point that scholars such as Diphoorn make is that situating the rise of private security in a context of 'state failure' is problematic. The view taken in this book is that it is better viewed in terms of the continuation and *reiteration* of more enduring power relations between social groups. It is in this regard that the book contributes to the literature on PSCs; it is not the agreement with others on the problematic notion of state failure, but the more overt placement of private security and other security actors in a socio-political context with a history that precedes these companies. This focus, as discussed in the following section, will be observed through the diverse security infrastructures and practices, which include PSCs and with which they often interact in particular ways. The aforementioned notion of *reiteration* is important. It will be repeated many times throughout this book and should be viewed as something between continuation and rupture. I take reiteration as an empirically apt framing of change in South Africa in recent decades. In a somewhat different way, some forms of *reiteration* are desirable for the purposes of progressive praxes, in accordance with notions of hegemonic rearticulation as described by Laclau and Mouffe (2014[1985]) and Laclau (1990).

Research approach: infrastructure, hegemony and space

Taking inspiration from Brian Larkin (2013) I define infrastructures as those resources that form a substrate for the circulation of ideas and practices, in this case, of crime fighting. These infrastructures are diverse and include traditional physical infrastructures such as roads as well as barriers of all types, from specially policed areas, often through the likes of closed-circuit television (CCTV), high walls and alarm systems. It also includes platforms through which divisive discourses and border-making are facilitated, such as social media, newspapers and routine activities, such as neighbourhood watches. Infrastructure also includes cognitive frames that fulfil the same functions as physical infrastructure in directing ideas and practices around crime fighting.

Private security companies and other security actors can influence the world they work in. In other words, they are at the same time also part of the substrate that influences security practices. An infrastructural approach does not dispute the discursive nature of security. However, the approach taken in this book, by focussing on the ideational, physical and agential aspects of discourse, also reveals more about the lived experiences of security actors mediated through physical and metaphysical structures. Moreover, compared to the nodal governance and network governance approaches, *the emphasis of the infrastructural approach is on flows*. What are the flows of ideas, people and objects that are facilitated and constrained through various security infrastructures? Although I have much respect for the body of work produced by an author such as Shearing, I do see the need for the relationships between security infrastructures to be relatively equal. Not all infrastructures are equal, nor are all citizens treated equally through the ideas and practices that circulate through security infrastructures. As such, there is a need for greater oversight in many instances. I have chosen a different post-structural slant compared to the IR scholars cited earlier, to elaborate the infrastructural approach described previously. By drawing on the Laclauian notion of hegemony (his work on this topic, initially in collaboration with Chantal Mouffe), I aim to produce constructive criticism and work towards preliminary suggestions for potentially productive praxes.

To better understand the aforementioned South African socio-political context and its hierarchies, I have chosen additional concepts to better frame and execute the study. The optic of hegemony, largely informed by the work of Ernesto Laclau (cf. Laclau and Mouffe, 2014 [1985]; Laclau, 2007) allows for more open-ended analyses when compared to the determinism of many other more radical progressive approaches. Firstly at the very foundation of Laclau's work is the radical contingency of the social order. Secondly, and significantly, there is a need to engage with, as opposed to alienating, diverse stakeholders. More overtly radical approaches potentially run the risk of such alienation.

Hegemony is both a dynamic political project and a form of governing (Howarth, 2010:310). In other words, hegemony is both 'an achieved (be it unstable) state' and a zone of contestation. Often oppressive government through this 'achieved' state occurs in a dynamic world, where change is constant, though not always fast. It is within this dynamism that opportunity for progressive change can be found. Laclau emphasises the incomplete nature of language and the fact that any social order is only a partially sutured precarious totality. When I speak of a social order, it is derived from Laclau and Mouffe's (2014[1985]) initial analysis. In this context, 'social order' should not to be confused with 'order' as something to be conserved through policing. Instead it refers to how the social is organised. 'Totalities' are constructed relationally through the contradictory logics of equivalence and difference. Certain logics of equivalence, that is, tenuously rendering diverse individuals and groups as equivalent, shores up a supposedly neat distinction between an inside and an outside. These equivalences may be undermined productively by the logic of difference, which calls out the artificial nature of the sutured social and by strategic usages of alternative equivalences to similarly challenge social orders. Claims to universal values and identities are always incomplete and open to challenge. As such, any notion of 'society' is inaccurate and should rather be replaced by the more generic notion of 'the social', as a space for continued contestation. Laclau (1990:8) notes that 'the social never manages to fully constitute itself as an objective order'. It is always dependent on an implied constitutive outside and as such the very political act of constitution can be undermined by highlighting its inverse and its oppressive consequences.

To this latter comment I wish to add that because of the limited resocialisation and changes in wealth distribution post-apartheid, hegemonic groups in South Africa are arguably more heterogeneous today than ever before. Herein lies a significant amount of potential for more productive alliances to be formed for the sake of security and as I will argue the greater social cohesion that will likely be required to deal with the problem of crime. The book will continuously highlight the lack of homogeneity amongst those associated with problematic identities and the lack of uniformity within individual subjectivities (Laclau and Mouffe, 2014[1985]:101) as something to be leveraged for a greater good. Progressives may claim and strategically redeploy logics of equivalence and difference. They may also highlight 'floating signifiers', indicative of the limitations and incompleteness of the aforementioned suturing, which circulate with the aid of diverse formal and informal infrastructures. Oppressive signifiers may be framed as inherent to the domination of some by others and as such in need of redefinition or realignment with more palatable signified concepts. This is where a radical and plural democracy is meant to enter, where a plurality of political praxes, based on the logics of difference and equivalence, develop. This form of democracy, based on continuously unfolding productive praxes, extends beyond elections and other formal institutions. Radical and plural democracy is anchored in the recognition of diverse struggles as equitable, and I would argue, in collaboration across an intertwined tapestry of political praxes. Mobilising alternative interpretations of a social order and a new positivity beyond the aforementioned deconstruction is a process Laclau and Mouffe call hegemonic rearticulation. Here the logic of opposition is joined by/to a logic of construction and there is a type of productive and enduring 'tension in openness' (Laclau and Mouffe, 2014 [1985]:174). Therefore, hegemony, according to the Laclauian perspective, is never overcome. There is also not necessarily a moment akin to a Jacobin revolution, unless it is a moment engaged in in full awareness of the radical contingency of the social.

Contra earlier Marxist readings of hegemony, most significantly by Gramsci (1971 edition), Laclau rejects the idea of a single centre in hegemony. The state is but one source of articulation and a conflict cannot be defined neatly between two antagonistic groups (Laclau and Mouffe, 2014[1985]:125). Instead, the radical contingency of the social, founded in the over-determination of all social identities, allows for numerous possible progressive praxes. These praxes, rather than a sudden rupture, may manifest as numerous reiterations. A Laclauian perspective is therefore far more concerned with *what is hegemonic* than it is with *who is hegemonic*. This does not mean that there are no hierarchies. Some subject positions are always more favourably placed. One consequence of this mode of thought, again contra traditional Marxists, is that categories such as class cannot be seen as a priori to the social. Radical and plural democracy implies diverse antagonisms within the social. Although, we may agree with this type of logic as a matter of principle, the South African case, with a history of racism, segregation and economic inequality, does call for these categories to feature strongly within the analysis. They are significant to the politics of crime. However, much like Laclau, I too challenge the assumption that the state is always the sole solution to social problems. Such a position, however, needs to be balanced with an awareness of the violence (broadly conceived) also effected by non-state security actors.

Given the spatial geography of apartheid and its post-apartheid reiteration, I wish to elaborate my analysis of hegemony in spatial terms with the aid of Henri Lefebvre's work, in particular his spatial triad and his notion of the right to the city. The latter implies equal opportunity for all to co-constitute the city and as Purcell (2013) argues, to 'live in it well'. The spatial triad can be linked to hegemony as follows. Lefebvre's distinction between space as a medium for social relations and a material product (cited in Gottdiener, 1993:132) mirrors the aforementioned distinction between hegemony as a zone of contestation and hegemony as a more or less unstable achieved state. Infrastructures direct the flow of routine activities and hegemonic practices. They are also the product of such practices. In particular, the analysis will interpret physical space, linked to the enduringly divisive and routinely reproduced, spatial geography of post-apartheid South Africa. Space is often interpreted through Lefebvre's (1991:1–67) triad of perceived space (physical space), spaces of representation (ideas, imaginations, visions and aspirations for remaking) and representations of space (more official representations such as maps and models associated with city planning). Lefebvre argues that there is a dialectic in the lived world between spaces of representation and representations of space (Keith and Pile, 1993:15). In this book I occasionally deploy these concepts in an overtly normative sense by drawing on the productive tension between the logics of equivalence and difference and to highlight potential avenues for plural democracy. These are strategic moves not meant to contradict the underlying ontology of radical contingency and the fundamental incompleteness of the social.

Contemporary attempts at (re-)representing space by officials are at times at loggerheads with exclusionary notions attached to the spaces of representation. As such, there is a link with McCann's (1999:132) argument that space is considered a key feature of racial identities and urban formations. These spaces shape discursive and other practices, but they are also, significantly, shaped by practice in the co-constitution of hegemony. The book also analyses virtual spaces, such as social media platforms and their relationship with physical and metaphorical space and as security infrastructures themselves. The language of exclusivity for example finds a platform online. These articulations correlate with (physical) spatial practices, such as the summoning by clients of PSCs to clear the streets of unwanted passers through.

Spatial practices can, however, still be disrupted by claiming contestable signifiers currently articulated by the members of the public and through their interactions with private security and deploying these reclaimed signifiers towards progressive ends. The same, I would argue, given the spatial dimensions of hegemony referred to here, holds true for spatial practices, which may be contested as oppressive or more subtly reshaped. By thinking with the notion of hegemony in post-structural terms, we may mitigate the perils of striving towards a definitive telos and as such it is at this point that the analysis diverges from Lefebvre's oeuvre. As has been noted, hegemony is not to be transcended. This is not possible. It can, however, be reframed, continuously, as a progressively more ethical and democratic space.

To summarise, the book analyses the security practices and related discourses facilitated through diverse security infrastructures, located in a particular South African socio-political context. The research is based on multiple qualitative methods deployed between 2016 and 2019. Particular relevant methodological details are shared in each chapter. I aim to identify problematic discourses and practices while highlighting the potential for change by applying logics of difference, opposition and, in the final chapter, logics of construction. This approach is preferred to typical approaches in CSS. It has the potential to move beyond a general notion of 'security as emancipation' (the Welsh School, cf. Booth, 1991), de-securitisation (the Copenhagen School, cf. Buzan et al., 1998) and the routine practices of securitisation (the PARIS School, cf. Bigo, 2014), by engaging with the sociality of security in an oppressive, yet unstable hegemonic context. This potentially allows for multiple points of entry possibly towards subtler, yet potentially meaningful, interventions for change. Stated differently, there is the very real problem of crime. This cannot be denied, especially in the South African context. In addition, however, there are divisive ancillary ideas that circulate and become attached to concerns over crime. I will aim to disentangle these two issues and highlight alternative understandings and social praxes, in the spirit of (a) more constructive post-apartheid South African project(s). The book is, therefore, a project that seeks to move beyond critique into the realm of practice, that is founded in the type of constructive criticism that Laclau's work enables. Moreover, if we are to view theory or research as political practice (praxis), then it is imperative that I attempt to be as empathetic as possible to the different actors mentioned in this work. The point is to identify possible avenues for collaboration and not to lose people along the way through arrogant, dismissive analyses. The emotive and pressing nature of the subject matter at hand does, however, likely make such alienation partially unavoidable. Sitting here in my study on the top floor of a block of flats, I am arguably as safe as one can be in South Africa. I will not tell people that their fears are unwarranted. I will rather try to come up with alternative ways of seeing particularly problematic and divisive ways of approaching the problem of crime.

Research site

Some crime scholars have complained about the lack of research focussed outside major cities (Rukus et al., 2018:1858). This book, unlike many of the studies mentioned earlier, is not based on research in a large metropolis. Potchefstroom is part of a small city in South Africa's North-West Province. It is located about 120 km South-West of Johannesburg. As part of the broader JB Marks Municipality, Potchefstroom's official population is 168,762 (StatsSA, 2017). The city is characterised by a spatial geography typical of South Africa. The Potchefstroom central business district (CBD) is surrounded by the affluent and middleclass suburbs of Baillie Park, Grimbeeck Park, Van der Hoff Park, the *Bult* (or Hill) and Dassierand. Potchefstroom is separated from a number of 'townships' by an industrial area. These townships mostly comprise informal settlements and generic structures built through the state housing scheme rolled out from the early 1990s, as part of the Reconstruction and Development Programme (RDP).[1] These settlements include Ikageng (population 87,701), which was previously designated for black Africans, Promosa (population 16125) and Mohadin (population

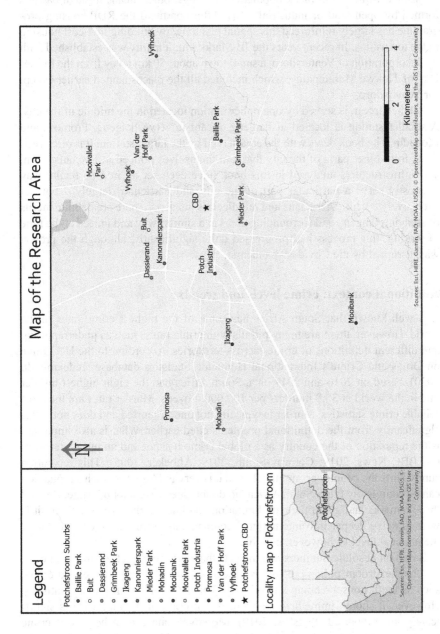

Map of the Research Area

Legend

Potchefstroom Suburbs

- Baillie Park
- Bult
- Dassierand
- Grimbeek Park
- Ikageng
- Kanonnierspark
- Mieder Park
- Mohadin
- Mooibank
- Mooivallei Park
- Potch Industria
- Promosa
- Van der Hoff Park
- Vyfhoek
- ★ Potchefstroom CBD

Locality map of Potchefstroom

Sources: Esri, HERE, Garmin, FAO, NOAA, USGS, © OpenStreetMap contributors, and the GIS User Community

Sources: Esri, HERE, Garmin, FAO, NOAA, USGS, © OpenStreetMap contributors, and the GIS User Community

Map 1.2 The research area

1601), which during apartheid were designated for 'coloured' (of mixed race) and 'Asian' South Africans respectively. Under apartheid the state actively tried to insulate whites from black populations amongst other means through barriers formed by open land or industrial areas. After apartheid the RDP housing programme has largely reinforced this spatial planning, by building low-cost housing in the townships. In recent years the JB Marks Municipality was established with the incorporation of Ventersdorp, a small town about 50 km away from the former City of Tlokwe Municipality, which included all the places named earlier except for Ventersdorp.

Potchefstroom is served by one police station located in the middle of the city. A smaller station is located in Ikageng, meant to serve Ikageng, Promosa and Mohadin. The book deals with the area served by the larger station. It is often residents from other parts of the city that find themselves policed in Potchefstroom. The infrastructures analysed in this book have been set up in a particular way to mostly serve a particular part of the city and particular, largely – but not exclusively – white, Afrikaans and middleclass interests. There are definite merits of studying Ikageng and surrounding areas in a similar way and indeed at the time of writing, this process had progressed quite significantly, although the project was disrupted by the Covid-19 pandemic.

Additional context: crime levels and trends

It is well known that South Africa has some of the highest crime rates in the world. However, there are many pitfalls with crime rates, such as underreporting and different definitions of crimes across territories. According to the UN Office on Drugs and Crime's International Homicide Statistics database (Indexmundi, 2020), based on 2016 and 2017 data, South Africa had the eight highest murder rate in the world at 35.9 murders per 100,000 citizens. Murder rates are the most reliable crime statistics. Murder is typically not underreported and does not suffer significantly from the definitional problems cited earlier. What is also important is the reputation of the country as a global crime hotspot and an unsafe territory (cf. BBC News, 2019; Conway-Smith, 2019; Altbeker, 2005). This reputation surely hurts the country economically and it is very unlikely that such a reputation can be completely false. South Africa, no doubt, does have a lot of crime. Beyond these remarks and given the data limitations outlined earlier, I am not sure it is wise to delve deeper into international comparisons. Some more comparison at the national level is, however, possible.

Based on absolute numbers and the acts that seem to attract the most concern, the Potchefstroom police precinct is especially prone to property-related crime, such as theft, housebreaking and robbery. This is borne out in official statistics. However, it should immediately be stated that crime in South Africa is significantly underreported (StatsSA, 2019). Moreover, one cannot be sure if crime is equally underreported everywhere. Therefore, Table 1.1, which compares the Potchefstroom, Wellington, Meyerton and Cape Town CBD precincts based on official statistics, should be viewed as a rough comparison only. The table is based on the 2018–2019 crime statistics, of selected crimes represented in absolute number of reported cases and per 100,000 residents for ease of comparison. The crimes selected are based on those most often reported in Potchefstroom, but it also includes murder as the most reliable statistic. The population serviced by each precinct is based on the 2011 Census. On account of urbanisation these populations have arguably grown. Moreover, the Cape Town Central Police Station, as the name suggests, is based in the CBD of a major metropole. That station is located in a relatively small CBD (for a metropole), with various apartment complexes, but essentially the smallest official population serviced by each of the four precincts. Because it is located in a CBD and it is not the only station for the larger urban unit, other rules of interpretation apply. Obviously many more people move in and out that precinct every day who may either perform crime or be victims of crime. I include that precinct here precisely because, in contrast with the other stations mentioned here, it is located in a major metropole. It therefore offers a different type of comparison. All three of the other stations share certain commonalities. Their populations are similar in size. They are all close to larger urban centres. These are Klerksdorp in the case of Potchefstroom, Paarl in the case of Wellington and Vereeniging in the case of Meyerton. In addition, all three of these non-metropolitan stations are about 40–60 minutes by car from a major metropole. These metropoles are Cape Town in the case of Wellington and Johannesburg in the case of Potchefstroom and Meyerton. It should also be noted that both Wellington and Potchefstroom are university towns. The North-West University's (NWU) main campus is located in Potchefstroom, along with other smaller higher education institutions. Potchefstroom also includes an agricultural college and a beautician academy. A satellite campus of the Cape Peninsula University of Technology (CPUT) is located in Wellington. The number of students in Potchefstroom are, however, far greater than Wellington. To reiterate, what follows is not in any way meant to be a comprehensive comparative analysis. The point is simply to illustrate that Potchefstroom does have relatively high levels of certain crimes and that concerns over crime cannot be dismissed easily.

All figures have been rounded up or down to two decimal points. The exact populations served by a particular station were taken from secondary sources. In three instances this was based on inferences by Geographical Information Systems (GIS) specialist, Adrian Frith provided on his personal website. All calculations are my own.

As the table indicates, only the Cape Town Central Police Station (hereafter CCPS) exceeds Potchefstroom in total amount of reported cases. With the exception of shoplifting, the same holds true for every other form of property crime and cases where property crime turns to contact crime, such as housebreaking when a resident is home. Wellington, however, has much higher levels of drug-related crime than any of the other smaller stations. Potchefstroom is second. This might be understandable as both settlements host institutions of higher learning. Interestingly, Meyerton generally has far less crime than the other stations. This town is a relatively blue-collar and lower-income area compared to Potchefstroom and Wellington. The case of Meyerton may therefore be partial evidence linking

Table 1.1 Comparative crime statistics

Population served Crime type	Potchefstroom 63,335+		Wellington 55,543++		Meyerton 55,283*		Cape Town Central 35,019**	
	N	Per 100,000	N	Per 100,000	N	Per 100,000	N	Per 100,000
Total crime	6,728	10,622.88	4,413	7,949.49	2,043	3,695.53	16,371	46,748.9
Murder	6	9.47	27	48.61	7	12.66	7	19.99
Sexual offences	65	102.63	74	133.23	65	117.58	56	159.91
Common robbery	209	329.99	61	109.82	10	18.09	1,070	3,055.48
Robbery with aggravating circumstances	315	497.36	158	284.46	139	251.43	718	2,050.32
Burglary at non-residential premises	190	299.99	232	417.69	96	173.65	243	693.91
Burglary at residential premises	861	1,359.44	478	860.59	310	560.75	515	1,470.63
Theft of motor vehicle and motorcycle	235	371.04	33	59.41	156	282.18	217	619.66
Theft out of or from motor vehicle	977	1,542.59	331	595.93	79	142.9	3,710	10,594.25
Drug-related crime	380	599.98	518	932.61	178	321.98	2,313	6,604.99
Shoplifting	97	153.15	92	165.64	24	43.41	766	2,187.38
Robbery at residential premises	37	58.42	11	19.8	47	85.02	17	48.55
Robbery at non-residential premises	9	14.21	12	21.6	23	41.6	29	82.81

+Source: https://census2011.adrianfrith.com/place/676004. Here I added rural areas of the previous City Tlokwe to the Potchefstroom population, as these are also served by the same station. I essentially subtracted the populations of Ikageng, Promosa and Mohadin from the entire City Tlokwe population as per the 2011 census. This might be a slight overestimation. However, if we were to use the source cited here, which was not developed for the same purposes as the current study, we would almost certainly underrepresent the population served by the Potchefstroom station quite substantially.

++Source: https://census2011.adrianfrith.com/place/166005, accessed June 2020

*Source: https://census2011.adrianfrith.com/place/761006, accessed June 2020

**Source: www.groundup.org.za/article/camps-bay-has-887-police-officers-100000-people-nyanga-has-161-why/, accessed June 2020

crime, especially property crime, to inequality on which I elaborate in Chapter 1. Potchefstroom, for example, might not be very unequal, but there is extreme poverty in the nearby and neighbouring Ikageng township.

While I conducted my research, there were a few particularly prominent crime concerns in Potchefstroom. One such concern was the so-called *Voordeurbende* (front door gang). They were named such on account of their particular modus operandi. The gang(s) would hit a particular part of the city, typically the wealthier suburbs, on Sunday mornings when many in this very religious community were in church or during weekdays, when people were at work. This gang were 'professional'. They would very quickly force open the front door and security gate, remove flat screen televisions, phones, laptop computers and jewellery and moved onto the next house. They left no tracks and removed themselves from the area, typically before the police or private security responded. This type of crime is prevalent not only in Potchefstroom but in middleclass suburbs throughout South Africa. When an occupant is accidentally found inside in the process of a housebreaking, the crime then escalates into another official category, a house robbery. These crimes are also quite common in Potchefstroom, but not nearly as common as housebreaking.

Theft from motor-vehicles and theft of motor-vehicles have also been very prevalent. The latter, at the time of writing, especially involved the very popular Volkswagen Polo. They are stolen, their tracking devices removed, often furnished with new number plates and driven, quickly, to the Gauteng Province from where they are apparently stripped into parts and sold to scrapyards or exported. Theft from vehicles entail broken windows or the remote jamming of central locking systems to gain access to vehicles when owners leave what they believed was a locked car. Here, typically, purses, laptops, phones and other valuables are taken, especially when they are visible from outside. Sometimes, perpetrators would even be so brazen as to briefly join the driver in their car, remove the valuables and calmly walk away, typically into a crowd. It is because of such instances that certain areas, particularly in the CBD, have been labelled as high crime zones, especially the area around the *Wandellaan* (Pedestrian Street) and nearby taxi rank, where black residents of Ikageng are dropped off and picked up by mini-bus taxi every day. It is also in this part of the city where pedestrian robberies are most common. The notion of stigmatised spaces in reference to this area amongst others is the focus of Chapter 3.

Another major concern, again much like the rest of South Africa, is illegal drugs. As is noted previously, Potchefstroom hosts a university campus, an agricultural college and a beautician academy. Students are a significant component of the clientele. By most accounts, a rather full range of substances are sold in the city, including more affordable substances such as Methcathinone, which is cheap and easy to produce. By most accounts illegal drugs are linked to West-African, often referred to as 'The Nigerians'. From this we must resist stereotyping deducing that all Nigerian or West-African nationals residing in South Africa are involved in the drug trade. The weight of evidence from my research, however, suggests that the trade in South Africa and in Potchefstroom overwhelmingly involves

Nigerian nationals. There are also rumours that the Nigerian groups are mere foot soldiers of South African residents living in the community. Police raids are reported at least annually. Individual dealers are caught, tried and even deported, but the drugs trade in Potchefstroom remains robust, possibly because kingpins have ways of creating distance between themselves and clients and the fact that there remains a demand for the product.

In addition to the aforementioned crime trends particular events, readily recalled, have shaped the popular consciousness. In 2014 for example the neighbourhood of *Oewersig* was raided by a heavily armed gang. Again, reportedly, residents gave the police coffee afterwards to 'calm them down', as South African Police Service (SAPS) members were in a state of shock after being significantly outgunned. The result of this watershed moment was the introduction of street cameras funded entirely by residents and monitored 24 hours a day by a local security company. We will revisit this development in due course as the Oewersig case study is the focus of Chapter 8. Another incident regularly recalled is the murder of a lady in Mooivalleipark (see Map 1.2) in the same year. She was shot dead in her house during a robbery one evening. This event is indicative of a particular tendency. The outskirts of the city are vulnerable on account of proximity to escape routes.

The structure of this book

The remainder of this book is divided into three parts, each of which include three chapters, and a concluding chapter. Part I comprising Chapters 1–3 builds on the theoretical remarks of the introductory chapter by conceptualising contemporary political economic and spatial arrangements in South Africa as an open frontier. Here the singular form is used, purely for the sake of analysis. It is a simplification, but one that is potentially analytically productive. Within this open frontier we find a non-continuous laager. The *laager* – a South African word for the proverb, 'to circle the wagons' – as a physical and metaphysical construct is the pivotal security infrastructure that materialises as a collective of other security infrastructures. The production and reiteration of the laager is the result of a symbiotic relationship between the physical and the ontological. Chapter 3 concludes Part I by explaining how the stigmatisation of particular outside spaces justifies the existence of the inside defined in through logics of equivalence. Part II comprising Chapters 4–6 deals with the echo chamber effect facilitated by the various security infrastructures through which ideas of crime and ancillary discourses circulate. Much legitimate knowledge of crime passes through the community policing forums (CPFs), the *Potchefstroom Herald* and PSC social media platforms in order to help citizens to protect themselves. In addition, however, these platforms, largely by way of content shared by the public, perpetuate problematic ideas associated with crime. Part III includes Chapters 7–9. This part deals more closely with the security routines of daily life. It focusses on the activities of various institutions, such as PSCs, the Oewersig Residents' Association and a newly formed Cachet Park City Improvement District (hereafter the CID). These chapters engage in depth with the daily routines of private or semi-private organisations in crime fighting. Part III also reflects upon the adverse consequences of these activities and, yet, the potential for progressive change that is found in daily routines. The concluding chapter considers the sutured hegemony and the daily flows of ideas and practices observed through the infrastructural lens. The chapter tentatively considers how strategies of opposition and construction directly and indirectly related to the politics of crime might look. Thereafter, it offers some comments on the contribution of the study to different bodies of literature.

Note

1　The RDP was the first iteration of economic policy after apartheid. This policy had a more social democratic slant when compared to the policies that succeeded it. It was soon replaced by the free market macro-economic strategy Growth Employment and Redistribution (GEAR) and associated policies from 1996. The RDP's housing scheme has however endured.

References

Abrahamsen, R. and Williams, M.C. 2009. Security beyond the state: Global security assemblages in international politics. *International Political Sociology*, 3(1), pp. 1–17.

Altbeker, A. 2005. Puzzling statistics: Is South Africa really the world's crime capital? *SA Crime Quarterly*, 11, pp. 1–8.

Atkinson, R. and Blandy, S. 2005. Introduction: International perspectives on the new enclavism and the rise of gated communities. *Housing Studies*, 20(2), pp. 177–186.

Avant, D.D. 2005. *The market for force: The consequences of privatizing security*. Cambridge, UK: Cambridge University Press.

Ballard, R. 2004. Assimilation, emigration, semigration and integration: White peoples Strategies for finding a comfort zone in post-apartheid South Africa, in N. Distiller and M. Steyn (eds.) *Under construction: Race and identity in South Africa today*. Johannesburg: Heineman, pp. 51–66.

BBC News. 2019. South Africa crime: Police figures show rising murder and sexual offences. Online: www.bbc.com/news/world-africa-49673944. Date of access: 27 February 2027.

Bigo, D. 2014. The (in)securitization practices of the three universes of EU border control: Military/Navy – border guards/police – database analysts. *Security Dialogue*, 45(3), pp. 209–225.

Bigo, D. 2008. International political sociology, in P.D. Williams (ed.) *Security studies: An introduction*. London: Routledge.

Booth, K. 1991. Security and emancipation. *Review of International Studies*, 17(4), pp. 313–326.

Buzan, B., Weaver, O. and de Wilde, J. 1998. *Security: A new framework for analysis*. Boulder: Lynne Rienner Publishers.

Conway-Smith, E. 2019. Soaring murder rate drives Cape Town up list of deadliest cities. *The Times*. 1 July. Online: www.thetimes.co.uk/article/soaring-murder-rate-drives-cape-town-up-list-of-deadliest-cities-xhqw3k0ld. Date of access: 27 February 2020.

Diphoorn, T.G. 2015a. The private security industry in urban management, in C. Hafer-burg and M. Huchzermeyer (eds.) *Urban governance in post-apartheid cities: Modes of engagement in South Africa's metropoles.* Pietermaritzburg: UKZN Press.

Diphoorn, T.G. 2015b. *Twilight policing private security and violence in urban South Africa.* Berkeley: University of California Press.

Du Pont, B. 2004. Security in the age of networks. *Policing and Society,* 14, pp. 76–91.

Gheciu, A. 2015. Commodifying security in the Balkans: Between liberal norms and illib-eral practices. *Journal of International Relations and Development,* 18, pp. 288–310

Gottdiener, M. 1993. A Marx for our time: Henri Lefebvre and the production of space. *Sociological Theory,* 11(1), pp. 129–134.

Gramsci, A. 1971 edition. *Selections from prison notebooks.* New York: International.

Howarth, D. 2010. Power, discourse, and policy: Articulating a hegemony approach to critical policy studies. *Critical Policy Studies,* 3(3–4), pp. 309–335.

Indexmundi. 2020. Intentional homicides per 100 000 people. Online: www.indexmundi. com/facts/indicators/VC.IHR.PSRC.P5/rankings. Date of access: 27 February 2020.

Jones, T and Newburn, T. 1999. Urban change and policing: Mass private property re-considered. *European Journal on Criminal Policy and Research,* 7(2), pp. 225–244.

Keith, M and Pile, S, 1993. Introduction. Part 2: The place of politics, in M. Keith and S. Pile (eds.) *Place and the politics of identity.* London: Routledge, pp. 22–40.

Laclau, E. 2007. *Emancipation(s).* London: Verso.

Laclau, E. 1990. *New reflections on the revolution of our time.* London: Verso.

Laclau, E. and Mouffe, C. 2014[1985]. *Hegemony and socialist strategy.* London: Verso.

Larkin, B. 2013. The politics and poetics on infrastructure. *Annual Review of Anthropol-ogy,* 42, pp. 327–343.

Lefebvre's, H. 1991. *The production of space.* Translated by Donald Nicholson-Smith. London: Blackwell.

Low, S. 2001. The edge and the center: Gated communities and the discourse of urban fear. *American Anthropologist,* 103, pp. 45–58.

McCann, E.J. 1999. Race, protest, and public space: Contextualizing Lefebvre in the U.S. city. *Antipode: A Radical Journal of Geography,* 31(2), pp. 163–184.

Percy, S. 2007. *Mercenaries: The history of a norm in international relations.* Oxford, UK: Oxford University Press.

Purcell, M. 2013. To inhabit well: Counterhegemonic movements and the right to the city. *Urban Geography,* 34(4), pp. 560–574.

Rukus, J., Warner, M.E. and Zhang, X. 2018. Community policing: Least effective where need is greatest. *Crime & Delinquency,* 64(14), pp. 1858–1881.

Shearing, C. and Stenning, P. 1981. Modern private security: Its growth and implications, in M. Tonry and N. Morris (eds.) *Crime and justice: An annual review of research.* Chi-cago: University of Chicago Press, pp. 193–246.

Shearing, C and Wood, J. 2003. Nodal governance, democracy and the new 'denizens'. *Journal of Law and Society,* 30(3), pp. 400–419.

Statistics South Africa (StasSA). 2019. Governance, public safety and justice survey 2018/19. Online: www.statssa.gov.za/?p=12620. Date of access: 27 February 2020.

Statistics South Africa (StatsSA). 2017. Tlokwe City council. Online: www.statssa.gov. za/?page_id=993&id=tlokwe-city-council-municipality. Date of access: 31 July 2017.

Part I

The reopened frontier and the fortified laager

I am attending a meeting in Ikageng as part of a Social Sciences Research Council (SSRC)/African Peacebuilding Network (APN) funded project, which was meant to follow on the project this book is primarily concerned with. The meeting I am at pertains to illegal squatters who built homes on public land. These residents have been frustrated by the state's delivery of houses under the otherwise defunct RDP of 1994. People have been waiting for houses for decades. Some have allegedly leapfrogged the official list because of connections within the state. The meeting is about a court order that has been passed against these illegal squatters. Execution of such orders, often by PSCs, typically entails breaking down informal dwellings. This often leaves people homeless. Those who called the meeting are trying to figure out the best way to fight the court order. I am mostly at sea as I do not speak Tswana, one of South Africa's 11 official languages. I only speak two and could not secure the services of an interpreter this time. It is Sunday. Then a member of the Socialist Revolutionary Workers Party of South Africa (SRWP) gets up. He is 'of Indian decent' as per South Africa's enduring perverse, racial classification system. He therefore speaks English. 'Comrades, why are we fighting for scraps. We should be occupying open land in Potchefstroom.' Then he says, 'there is a railway line that runs North to South through this municipality. On this side, we are black. On that side it is the white middleclass, with only a sprinkling of the new black middleclass. The people on that side do not care about us.' I am the only white person in the hall. I should perhaps be more uncomfortable, but I am not. Through all my years of doing research in townships and informal settlements, I have never felt unsafe. It is difficult to argue with this SRWP member's reasoning. Of course, there might be more nuance to the matter, but nuance and mobilisation are not always, practically speaking, the best bedfellows.

*****

The experience mentioned here represents much of what the first part of this book is about. South Africa still has internal borders. Only now, border-making is less formal. Popular mobilisation takes place in South Africa frequently. Its effect on progressive change has, however, been somewhat limited. The people on the 'other'/wrong side of the railway lines often break the law, be it through illegal occupation of public land or through criminal activity. This sometimes happens in

DOI: 10.4324/9781003028185-2

Potchefstroom, but it is not because of some innate characteristics. Such a notion is as repugnant as it is ludicrous. And yes, the railway line does not provide the perfectly neat border implied earlier. The reality is a bit more complex. As the example given here suggests, race remains a major problem in contemporary South Africa, but there is a new emerging black middle class and in places even a black upper class. Within the context briefly outlined here Part I, which comprises three chapters, does two things.

Firstly, Part I provides a macro-level analysis which creates a context for contemporary security infrastructures in Potchefstroom. Secondly, as a set of chapters mostly about border-making, it introduces concepts that may be of use elsewhere. In the process it does not draw on Border Studies in depth. Instead I try to maintain the focus on infrastructure and hegemony and juxtapose key concepts only to literature on border-making, as it pertains to crime and supposedly remedial infrastructures. The first chapter conceives of a contemporary 'reopened frontier' that has unfolded for the past 50 years, at least. In this context a political economic strategy of closing the frontier through the biopolitical abandonment of the majority of citizens is futile and paradoxical and confronted with the adversities of its internal contradictions. This logical is paired with the spatial practice of frontier governance, which alleviates some of the major contradictions of the former logic. There is, however, no closing of the frontier. Instead the ontological structure of the frontier should be abandoned and replaced with real integration. The second chapter theorises the sum total of physical and ontological security infrastructures as a laager, the physical and cognitive manifestation of the idiom 'to circle the wagons'. Importantly, those inside the laager, and by extension the ideas and practices of the current hegemonic order, are more diverse than ever before. As such, the laager, although a South African word historically associated with white and Afrikaans South Africans, is compared to notions of 'semigration' (Ballard, 2004) and fortresses (Blakely and Snyder, 1997) as a concept that might be of use in other contexts, removed from the more archetypal example of Potchefstroom, the South African setting and from gated communities. If we follow a Laclauian logic, any imagined interior also has an important exterior on which it is silent. This exterior is an amalgam or equivalence of various differential relations. Such a constitutive outside is dealt with in Chapter 3. That chapter argues that this dependence pertains to how crime ridden, stigmatised, spaces have the inadvertent effect of justifying the laager or interior and associated exclusionary practices. These justifications are various and strengthen the laager and contemporary, though untenable, processes of biopolitical abandonment.

This first part of the book might seem esoteric to some more 'practically inclined' commentators. It is, however, indispensable for our understanding of the collective flows in the form of everyday security practices and associated ideas that circulate through contemporary security infrastructures. The laager and the political economy which underlies it is pivotal to any infrastructural analysis of the politics of crime and security in Potchefstroom and possibly elsewhere. It is pivotal as a collective of nodal points in the form of diverse security infrastructures, from the physical to the metaphysical.

References

Ballard, R. 2004. Assimilation, emigration, semigration and integration: White peoples' strategies for finding a comfort zone in post-apartheid South Africa, in N. Distiller and M. Steyn (eds.) *Under construction: Race and identity in South Africa today.* Sandton: Heinemann, pp. 51–66.

Blakely, E. and Snyder, M. 1997. *Fortress America: Gated communities in the United States.* Washington, DC: Brookings Institution Press.

1 Biopolitics and the reopened frontier

Introduction

This chapter describes the forces that have potentiated and at times called for security infrastructures. To explain this, we must go back a few hundred years to the 19th century. But, in doing so, it should be noted that this venture into history is largely at the level of metaphor and as an analogy of prevailing logics of collective security governance. It should also be noted that I am not the first to view the problematic of crime in contemporary South Africa through the lens of a frontier. Dieltiens (2011:96) in the following quotations outlines much of the significance of the frontier to the politics of crime.

> There is no clearer sign of the existing frontier in South Africa than the security cordons around real estate developments. The coils of razor wire and palisade fencing mark a boundary around privilege, guarded by private security companies against a surprise attack.
>
> If the abiding imperative of frontier law-making is the protection of property and the creation of conditions and the subjectivity for future accumulation, then the punishment of crime retains its political and economic utility.
>
> (Dieltiens, 2011:96)

In the 1800s Western expansion from the Cape Colony primarily occurred on two fronts, the so-called eastern and northern frontiers. Frontiers can exist in two states opened and closing. A completely closed frontier is no longer a frontier. The frontier seizes to exist once a relatively stable, be it dynamic, hegemony has been achieved in a given territory. This hegemony is not stable because it is static. It is stable because it is either enforced or because there is significant acceptance of the rules whereby everyday life is practised. The argument presented in this chapter is that the unravelling of apartheid has heralded in a somewhat reopened frontier. One might speak of various frontiers, outside the gates of a security estate, outside the bounds of a policed neighbourhood or city improvement district. However, for the sake of argument, I will employ the singular frontier. The frontier remains open, though there continues to be attempts to close it. In other words, there are various attempts to instil a set of practices that can stabilise the current order. The

DOI: 10.4324/9781003028185-3

problem, however, remains that there is widespread dissonance throughout the South African state and social order.

Since the early 1970s attempts at closing the frontier have been associated with the biopolitical abandonment of millions of residents. As such, we may speak of 'closure through abandonment', as a prevailing logic. Closure through abandonment is a political economic technique that seeks to stabilise a social order that prioritises some forms of life through maximal accumulation, often at the expense of others. This technique is continuously confronted by its internal contradictions, and as such, its failures. In practice, therefore, closure through abandonment is augmented by another set of strategies, frontier governance. Frontier governance is a set of spatial practices that draws on security infrastructures to create physical barriers between the groups polarised by way of closure through abandonment. Frontier governance involves diverse and complex forms of border-making.

In addition to briefly revisiting the frontier historiography on South Africa and explaining the notion of a reopened frontier, the chapter will also deal with biopolitics. In particular, I will draw on the notion of abandonment, typically of those deemed superfluous to dominant ways of life and the needs of capital. This happens in the context of an economy with high levels of financialisation but with low levels of industrialisation and thus low levels of employment for an inappropriately skilled workforce. Attempts at closing the frontier through abandonment has created a market for diverse technologies and services to mitigate fear. In this stabilisation of the current order, private security is often instrumental. This way of dealing with the problem is, however, arguably self-contradictory and therefore threatens the very sustainability of the polity. After explaining the nature and significance of these security infrastructures the chapter concludes and the argument is carried forward throughout the book, that alternative approaches are required to a logic of closing the frontier. We may hopefully rearticulate the hegemonic order to dispense with the notion of a frontier instead of seeking to close it. For now, as a heuristic device, the notion of an open frontier, in my view, is strategically appealing. By viewing the contemporary polity as such, one may recognise the relationship between stratification (on various grounds) and space. In addition, through a Laclauian lens, we may contemplate ways of disrupting frontier governance, thus not so much closing it, as heralding in a new common sense that sees less purpose for an oppressive frontier and the modes of thought that accompany it. The frontier image also allows us to consider the spatial dynamics of crime, security and politics in South Africa by drawing on the particular case of Potchefstroom. This book cannot possibly provide a blue print for a new dispensation. The radical contingency of the social requires multiple, continuous and nimble progressive politics.

This literature-based chapter further develops the notion of a reopened frontier, the logic of closure through abandonment and the associated logic of frontier governance. Thereafter it offers an explanation for some (especially property related) crimes in South Africa by bringing the contradictions of closure through abandonment into conversation with the concept of multidimensional poverty and Strain Theory.

The reopened frontier

The notion of a frontier is a useful prism through which colonial encounters can be viewed. Typically, this approach is applied to initial encounters up to the point where a central political authority is established. The analogy of a reopened frontier is perhaps a bit strange. I do, however, believe it is a meaningful way of looking at contemporary South Africa that allows us to consider potentially productive ways of overcoming physical and discursive manifestations of frontier governance. Consequently, it will become apparent how we may more comprehensively intervene upon contemporary problems such as crime. There is a difference between South Africa over the past 50 years and the same territory in the 1800s. What became modern-day South Africa has gone through various colonial dispensations and then formal apartheid, the end of which heralded in an imperfect democratic order. Contemporary 'encounters' are between two ideal-type elites and the majority of the populace, which remains impoverished and mostly black. Koelble (2018:5), drawing on the seminal work of Partha Chatterjee, calls this majority political society. They make up well over 60% of the population and live on the fringes of society, both spatially and economically. This group is almost entirely decoupled from the formal economic system and they are highly reliant on welfare. It should be understood that this reliance of welfare is not in terms of direct transfers to the unemployed, but instead based on the fact that up to three generations may share a single state-sponsored old-aged grant of around USD115 per month.[1] This is sometimes augmented by modest child support grants and/or disability grants.

The two elites in question are ideal-types in the Weberian sense. They are the economic elite, associated with historical legacies of cumulative wealth and opportunity. This elite own, manage or work in major business operations and they are relatively privileged, just like their parents and parents' parents were. They are mostly white. It should also be made clear that not all white South Africans are or ever were middleclass or wealthy. The second group, the political elite are the beneficiaries of recent upward social mobility. They have been favourably placed in regards to the negotiations for a democratic South African and to recent broad-based black economic empowerment (BBBEE or BEE for short) initiatives.[2] Many have benefitted from opportunities such as scholarships. A small proportion of this elite has been implicated in corruption and ill-gotten gains, but the majority are not. Many have worked extremely hard to use the new opportunities that only some have had access to since the advent of democracy. Many of these South Africans also pay 'black tax'. As the first person in a family to earn a decent salary and often have to support an extended network of people. In terms of the topic of this book, the spatial composition of both elites is predominantly in the fortified enclaves that will be described throughout this book. However, some within the political elite are not capable of living in those areas, likely because of black tax. This implies, hierarchies, also within and outside of the elite enclaves, but space does not allow us to do justice to that topic in this book.

Legassick's notion of a frontier zone is useful for the current analysis. Such a zone is an area where different societies clash, but where none of them have political hegemony (Legassick in Rasool, 2006:25). To be clear, the hegemony Legassick is referring to is decidedly Marxist and not post-structural. His point is that neither grouping is politically dominant in the frontier zone. The frontier has traditionally been conceived of as a space on the margins of colonial government. Colonial society was expanding, yet those areas beyond colonial rule were still contested and as such open. Eventually, a sense of permanence was reached once all contested areas were incorporated into a single territory and subjected to hegemonic government.

The frontier tradition in South African historiography can be divided into three broad traditions. They are the *Afrikaner*[3] nationalist, liberal and revisionist schools. The Afrikaner nationalist historiography was generally concerned with the story of the *Afrikaner* in conflict with British imperialism and 'black barbarism'. The *trekboere* (pastoral farmers) were viewed as pioneers carrying civilisation and Christianity inland in their migration from the Cape Colony (Visser, 2004:4–5). The frontier was therefore a place of confrontation with supposedly inferior savages in a quest to gain independence from the oppressive British Empire governing from Cape Town. The frontier was also where a new national identity emerged. The Liberal School rightly viewed the Afrikaner historiography as racist. In contrast to the largely descriptive approach of the latter the liberal historians took a more analytical approach and viewed all relevant actors as part of a single South African economy and society. A major argument from the liberal historians was that the frontier was where racism was cultivated and then carried inland, eventually to be entrenched in the laws of the *Zuid-Afrikaasche Republiek* (ZAR South African Republic) and Orange Free State Republic. These republics were formed by Afrikaners in central South Africa. The ZAR was initially governed from Potchefstroom and then Pretoria. The Orange Free State was governed from Bloemfontein. The frontier was where trekboere would for the first time reflect upon themselves as white in Africa. In the process notions of superiority to native residents emerged (Walker, 1928).

The Revisionist School challenged some of the liberal school's findings. Informed by Marxism, their focus was largely on the incorporation of new territories and populations into the exploitative processes of South African and international capitalism. Largely independent pre-capitalist black populations were proletarianised as urban wage labourers (Visser, 2004:10). Frontier zones were products of colonising societies in the process of accumulating and rendering new territories useful (cf. Leggassick, 1980). The frontier was therefore the precursor to the class based and exploitative migrant labour system on which Herold Wolpe (1972) produced a seminal work. The discovery and mining of minerals in South Africa was significant in this process as according to Wolpe black South Africans were incorporated into these industries at below subsistence wage levels. These wages were subsidised by the pre-capitalist economy in the rural areas. The frontiers in southern Africa were initially points of contact between white stock farmers searching for new land and African livestock owners. In general,

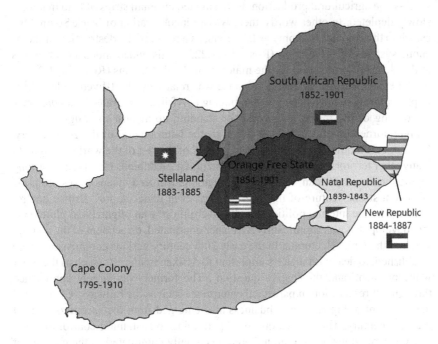

Map 1.3 South African republics, pre-unification

Source: Adapted from https://upload.wikimedia.org/wikipedia/commons/thumb/5/57/Boer_
republics_nl.svg/1024px-Boer_republics_nl.svg.png

the colonists fought the Africans for grazing land and livestock (Ross, 1981:212). Eventually the two *Boer*[4] republics, the Orange Free State and the ZAR,[5] were formed. However, with the discovery of gold and diamonds in Johannesburg and Kimberly respectively, the British became interested and conflict broke out in the 1890s over these precious resources.

My concern is largely with the Revisionist School, because it helps us understand the hierarchical structures of South African society throughout the 20th century. The revisionist view is the closest of the three perspectives to the one employed in this book. However, it should be clear from the introductory chapter, vast parts of the current chapter and the remainder of the book, that classical Marxism is not the position from which I am arguing. One may argue that the frontiers were 'closed' through proletarianisation *en mass* and the invocation of racial hierarchies from the later decades of the 1800s through the South African War of 1899–1902,[6] the creation of the Union of South Africa, as a self-governing British territory and the 1913 Land Act. The latter would eventually, once it was fully realised, mean that 87% of the population who were black could only own 13% of South African land. Many men were therefore forced to work precarious low wage jobs, typically on the mines, while rural homesteads, as a source of small-scale agricultural production, were situated on small strips of land that were slowly depleted. In other words, there was an incorporation of black South Africans into the colonial economy as labourers, as a result of the destruction of indigenous systems of production (Ross, 1981:223). This ended independent access to the means of production for the majority of South Africans (Ross, 1981:222).

The era of *de facto* British dominance was relatively short-lived, although its impact appears to have been quite damaging to Afrikaner morale. This *Volk*, once the darling of the rest of the world for standing up against the mighty British Empire, turned to a fully-fledged pariah by the latter quarter of the 20th century. Of course, it should be noted that this 'darling' of the rest of the world, along with many other territories across the globe, also practised dubious racial relations long before the South African War. Afrikaner nationalism as a means to restore pride became a strong feature of the 20th century, even under *de jure* British control until 1961. The internal politics of self-governing saw an oligarchical plural system develop that was significantly Afrikaner dominated, on account of the composition of the white electorate. In the early 20th century various commissions were established, to deal with matters important to Afrikaner interests, such as the 'poor white problem' and the 'native question'. The former was largely approached through job reservation in parastatal enterprises such as the railways. Clandestine development of big Afrikaner business also emerged from the second quarter of the 20th century. The 'native question' pertaining to sustaining control of a territory where whites were outnumbered, eventually culminated in the election of the National Party (NP) under DF Malan in 1948 and their policy of apartheid. Although elements of apartheid (directly translated as separateness) as a practice was already commonplace, apartheid as official policy was more formalised and comprehensive, including 'petty' and 'grand' apartheid. Petty apartheid related to practices such as different entrances to stores for whites and blacks, while grand apartheid had more to do with the broader policies regulating the movement and intermingling of people. This included the pass laws that limited the conditions by which blacks would be allowed in white areas. Grand apartheid also included the *Immorality Act* of 1927, repealed and reintroduced in 1955 and *Prohibition of Mixed Marriages Act* of 1949 that were meant to prevent sexual intercourse and marriages between different races. Very significant of course was the *Group Areas Act of 1950*. It determined who could live and own land where. For fuller elucidation of these and other apartheid policies see Van Riet (2014, in particular chapter 4). The NP would remain in power until 1994, although as I will argue its control over society started to wane long before that.

Closure through biopolitical abandonment

Biopolitics concerns the administration and regulation of life at the aggregate level. In the sense envisaged by Foucault (2008), it concerns the entry of aspects of human life into the realm of knowledge and power and associated political techniques. As such, biopolitics is the government of individuals and collectives through practices of correction, normalisation, disciplining, therapeutics, optimisation and exclusion (Lemke, 2011:5). This book is largely focussed on the latter, exclusion or abandonment, and the implications thereof. By 'making live and letting die' biopower is different from sovereign power. Sovereign power 'lets live and takes life'. Biopower fosters life or disallows it to the point of death. This takes place by disciplining the individual and regulating the general population (Lemke, 2011:36). Different norms regulate and discipline citizens, often indirectly, through their experiences and forms of acceptable behaviour, and, consequently, by shaping their frames of reference. Contemporary forms of biopolitics are typically strongly correlated to liberalism and its many guises. There have been previous forms biopolitical abandonment more similar to sovereign power. Hannah Arendt highlighted these in the 1950s, following her experiences in Germany during the Second World War (Arendt, 1953). To my mind, the abandonment of many black South Africans under apartheid resonates in part with Arendt's account of how Germany dealt with 'superfluous' groups, but only in the sense that it more forcibly regulated the movement of black South Africans. Apartheid did not include mass exterminations. A more Foucauldian understanding of biopolitics, linked to liberalism, helps to explain the abandonment that has occurred since the end of apartheid.

As an aside, some readers might ask why I have chosen a Foucauldian understanding of biopolitical abandonment, instead of the more recent work of Giorgio Agamben (1995). The answer to this question, without taking up too much space, lies in Agamben's focus on sovereign power, which distinguishes between citizenship and political life on the one hand and the state of exception and bare life on the other. A Laclauian notion of hegemony does not necessarily abandon the notion of sovereignty, but it certainly renders it far more complicated, as the state is but one source of hegemonic articulation amongst many. Even more importantly, in accordance with the logic of this book, the fact that Agamben leaves us with two somewhat bounded categories of people and life, *Zoé* (political existence) and *bios* (bare life), is far too limiting. This book informed by the aforementioned Laclauian notion of hegemony attempts to understand the diverse and non-unitary subjects involved in the response to crime. Later chapters will reveal that these actors and the categories they belong to are actually quite complex and heterogeneous.

In South Africa, the frontier reopened and biopolitical abandonment strategies were employed in self-contradictory attempts to close it. These abandonment strategies changed over the years. The initial unravelling of the system of apartheid was accompanied by strategies for preserving the regime. This was often in contradiction to the wishes of big business. Arendt's notion of superfluity refers to groups deemed weak and inferior and unfit to live. Totalitarian regimes, she argues often seek to rid the world of these 'dying' classes, hastening what is deemed 'inevitable' (Arendt, 1953:313, 318). We may of course argue convincingly that apartheid South Africa was authoritarian and not totalitarian. Given the potentially emotive nature of this discussion and its peripheral relation to my point, I would prefer not to pause on this matter. My point is simply that abandonment under apartheid was more like the abandonment described by Arendt and less like that described by Foucault and by extension Povinelli (2011) who I will draw upon in the following. Abandonment during apartheid was a more overt directive from above in a largely hierarchical system. This changed when apartheid ended. In democratic South Africa, power has been far more dispersed through society and the Foucauldian interpretation of governing discourses at the aggregate level, enacted through an array of actors, offer a far more plausible explanation.

The frontier 'reopened' by the mid the 1970s. The initial phase of Import Substitution Industrialisation (ISI) had run its course. It had produced growth, but the limited (largely white) market was too small for ISI to expand further. The government had a choice to make. They could either persist with apartheid or persist with the diversification of the economy (cf. Schneider, 2000:419–420; Liu and Saal, 1999:17). The former was chosen for the time being, until the economic track the country embarked upon, combined with domestic and foreign political pressure and, not to mention changes in geopolitics, gradually made apartheid untenable. Domestic tensions were consistent from the early to mid-70s, but there were a few key moments that may be emphasised here. The Durban strikes of 1973 was significant as it heralded in an era of an increasing influence of black trade unions in South African politics (Buhlungu, 2009:93). The 1976 Soweto uprising of students who refused Bantu Education and tuition in Afrikaans was violently oppressed.[7] This key date in history arguably stimulated further internal and external sentiments against the apartheid state. Two successive states of emergency followed in the 1980s. The state was fighting the black population in the townships, which coincidentally created a need for PSCs to do some of the work of the South African Police (SAP, as it was known then) in white areas and to guard national key points. In 1985 president Botha gave his infamous 'Rubicon Speech', where he told the rest of the world to 'do their damndest'. The West, for long an ally of the apartheid regime, obliged, a move like enabled by shifting geopolitical dynamics. Chase Manhattan was the first major Western institution to refuse doing business with South Africa. Others, such as Barclays Bank followed. During the 1980s the gold price declined. Military expenditure was high and the move from ISI to export led growth, which required subsidies, was too costly in light of increased sanctions (Liu and Saal, 1999:2). Then with the end of the Cold War, the apartheid regime and liberation movements reached a mutually hurting stalemate (cf. Van Riet, 2016:102–103). The NP government had lost their benefactors, so too did the African National Congress (ANC) and other liberation movements supported by the USSR. Negotiations followed, although these did not always happen in good faith. The so-called third force in the form of seemingly state sponsored black on black violence, was, it would appear, meant to weaken the ANC's position at the negotiation table (cf. Ellis, 1998:287). At the same time the far right mobilised and enjoyed more support than before. Seemingly, the ferocity of this support was dealt a final blow with the 'battle of Bophuthatswana', where three *Afrikaner Weerstandsbeweging* (AWB, Afrikaner Resistance Movement) men were executed live on international television.[8,9]

A settlement followed protracted negotiations during the early 1990s and culminated in the country's first democratic elections on 27 April 1994. The terms of this settlement has increasingly become a source of discord in contemporary South Africa. There is a body of literature that developed from the 1990s that deals with the so-called 'selling out' of the revolution (cf. Bond, 2005; Terreblanche, 2002; Marais, 1998). The consensus from this literature is more or less that the settlement set up a situation where much of the privilege of white South Africans remained intact while a new black middleclass grew largely through BEE initiatives. Today, in absolute numbers, this black middleclass exceeds the size of the white middleclass, but relative to each population group's size, the black middleclass remains relatively small. Black economic empowerment has not been as broad based as the name of the official policy might suggest. As the argument goes, this is because of the influence of the 'end of history' narrative and neoliberal paradigms immediately after apartheid. Growing poverty and inequality after apartheid has largely been linked to macroeconomic policies instituted under president Mbeki.

When former president Mbeki was ousted by Jacob Zuma as leader of the ANC, it has now been revealed large-scale corruption intensified, through continued ANC cadre deployment to local municipalities and the phenomenon of state capture (Desai, 2018). There are many theories as to why Zuma eventually replaced Mbeki as the leader of the ANC. These range from a view that Mbeki's economic policies were too conservative, that he was too dictatorial or that there was a 'coalition of the aggrieved' within the ANC (Southall, 2009:325). This implied a battle between those who benefitted from government tenders under Mbeki and those left out (Southall, 2009:327). Meanwhile Zuma was a very charismatic and fluid character who was able to court all kinds and establish a broad support base. I believe that two things are clear. The politics of patronage did not start after Mbeki. Also, economic policies under Mbeki (and after Mbeki too) have not managed to win over the majority of citizens, even though the ANC is still a dominant party at national level and in most provinces and municipalities. Cadre deployment started in the 1990s as a means of transforming the bureaucracy which was 95% white. Cadre deployment entails that bureaucracies are politicised through centralised deployments of party members. As such, bureaucrats are beholden onto those who deploy them first and foremost and not the public they are meant to serve. This practice has contributed to perverse patron-client relations and the near collapse of many basic functions in the local governmental sphere (cf. Thebe, 2017; Franks, 2014).

Elizabeth Povinelli (2011) develops a notion of abandonment as the absence of state care during what she calls late liberalism. Late liberalism is a period characterised firstly by wide spread neoliberalism and secondly by failing multicultural policies. Both aspects have been a major part of post-apartheid South Africa. Multiculturalism through constitutional design sought to walk a tightrope between individual and group rights. Neoliberal policies have permeated most of the post-apartheid years, since the inception of the Growth Employment and Redistribution (GEAR) macroeconomic strategy in 1996. Neoliberalism, or some might find the term 'globalisation' more palatable, is a contested concept. There can be little doubt that the term has often taken on the form of an empty signifier. It is a tricky concept too, in the sense that we arguably cannot do without it. Like most other concepts, the responsible author should explain what it means within the context it is used. In the case of South Africa, neoliberalism has included at least three elements: (1) limited barriers to international trade, (2) stable interest rates to keep inflation within a predetermined range and (3) hyper-marketisation. South Africa has a very open economy (Koelble, 2018:4). Because of the highly financialised structure of the economy, this openness causes much economic instability. Recent decades have seen portfolio investment being significantly substituted for foreign direct investment (FDI). Hyper-marketisation includes rendering new things as subject to market mechanisms. This has included increased regulation and outsourcing of public goods through market mechanisms. Although the South African state has arguably not shrunk significantly as has been the case elsewhere, the state has become increasingly beholden onto a type of governmentality whereby knowledge is deemed to exist primarily outside of the state. As such, this knowledge should be consulted. In many instances, this has created a barrier of sorts, of consultants, between the state and local level accountability. Moreover, consultants are not the state and also escape meaningful accountability.

Through the logic of GEAR stable monetary policy and obedience to dominant free trade dogma would apparently attract investment. This would fuel growth, employment and redistribution. The strategy seemed ill-conceived by many at the time and subsequently, even by president Mbeki's own brother, who is an economist. Moeletsi Goduka Mbeki (cited in Southall, 2009:327) has argued that GEAR and BEE did nothing to change the underlying apartheid economy. Official policy used what Povinelli refers to as 'bracketing', that is, putting on hold the ethical obligation to alleviate suffering in the hope that the problem will resolve itself or at least become less severe. Thus 'bracketing' in the South African case seems much like the more familiar trickle-down economics based on redistribution via growth. Bracketing therefore is consistent with the political economic logic of closure through abandonment introduced in this chapter. The results, as is common knowledge now, have been quite the opposite of redistribution. Inequality has grown. Moreover, these adverse consequences were exacerbated by the apparent lack of direction and grand corruption during the Zuma years. The Accelerated and Shared Growth Initiative for South Africa (ASGISA), that was meant to render more inclusive growth, was ended by Zuma in 2009 instead of 2014, as initially intended. The New Growth Path (NGP) developed by the Zuma administration and the National Development Plan 2030 (NDP) is yet to gain significant traction. Adopted in 2012 the latter has remained largely unimplemented and as Swilling (2019:25) argues it is flawed in any case, in terms of its ability to arrest continued marginality.

> the National Development Plan (NDP) is pessimistic about the future of manufacturing, saying virtually nothing about de-financialisation, and is vague when it comes to achieving employment-centred development in an environment where trade unions have policy influence.

Corruption took on new levels as state-owned enterprises became the targets of high-level patronage networks. The rhetoric of 'radical economic transformation' was spun along with 'white monopoly capital' as its apparent foe. Meanwhile, the contracts awarded through state-owned enterprises (SOEs) did not benefit the marginalised black population (Desai, 2018:502). It was a small elite drawing on the established liberal practice of buying in expertise, and radical rhetoric, that acted simultaneously as camouflage for looting. As Desai (2018) argues, this is money that could have been used for social projects, which as I will argue is an important part of crime reduction. The economy has continued to deindustrialised in recent decades. There can be little doubt that liberal trade regimes and labour practices in the East, as well as a vibrant trade union sector in South Africa, has much to do with this. It is, however, unfortunate that the ANC government could not come up with and follow through with a plan by which to deal with the adverse consequences of cheap Asian labour. The country's economy has a largely hollowed-out structure, as is mentioned earlier by Swilling. There is a large primary sector and a large tertiary sector, the latter significantly in financial services. Unemployment, in the narrowest sense, only including those of working age 'who have actively looked for employment in the formal sector during the past two weeks', is at 29.1% (StatsSA, 2020). The broader definition, including those employed in the informal sector and discouraged job seekers, but excluding those of working age who are studying, places unemployment at 40.1%. Youth unemployment in 2018 was hovering around 50% (Desai, 2018:510). The Covid-19 pandemic has likely made this dire situation even worse. Millions of people have lost their jobs during the lockdown period of 2020 (NIDS/CRAM, 2020:3).

As the previous discussion suggests, abandonment in the post-apartheid polity works by way of government through prevalent discourses. Self-sufficiency and free markets, as a selection of neoliberal practices, alongside radical discourse, certainly form part of the mix, but so too do discourses of fear and the division between those who require protection and those from whom protection is required. The latter are often those deemed too 'lazy' or 'unwilling' to be self-reliant. These types of attitudes by many of the more affluent members of society is quite apparent in some of the chapters that follow. Of course, none of the discussions in this chapter means that large-scale abandonment did not occur before the early 1970s. White and especially Afrikaner dominance was more or less in tact before 1970. As the contradictions of the status quo became apparent, the response has been more abandonment as opposed to drawing more citizens into the social and economic order. Some might argue that the system was unstable before the 1970s. I would not disagree, but, my argument relates to when this instability became apparent and the metaphorical frontier, in a sense, began to reopen.

Some might argue that the post-apartheid state is paying significant amounts of money on care in the form of state old-aged grants, child support grants and disability grants and that a proposed universal income grant would increase the welfare bill even more. South Africa indeed does have the largest social security system in Africa. While these grants often tend to benefit more people than merely the official recipient, the grants are still relatively low and have up to now not been targeted at working aged adults. Moreover, one could argue that these grants represent the bare minimum needed to ensure relative stability in the polity. They therefore serve the logics of closure through abandonment and frontier governance by way of stabilising or lending a semblance of legitimacy to an unjust social order. Consider the following perspective by Koelble (2018:20):

> Political society gets what it is demanding – some form of governmentality and the resulting access to some resources such as infrastructure – but it gets that in very measured, minimal doses. And the political/economic elite gets away with the lion's share of the commodity bonanza.

This quotation hints at what makes South Africa different from many other African and developing countries. The relative wealth of a small minority exists side by side with the intergenerational transfer of marginality. Conspicuous consumption exists as a depraved reminder of the failures of democratic transition. Through regular and widespread protest action, many are voicing their discontent with the failure of the state to deliver what is expected. The severe inequality and moral bankruptcy of the status quo is a point to be raised again in the following, where I attempt to explain crime in South Africa and its relationship to inequality.

With democracy there also came a new spatial politics. Apartheid was over and everyone was legally afforded the right of freedom of movement, and the right to live and own property wherever they like. These are *de jure* rights. Not everyone is able to afford such rights. With this spatial politics, white fear and to some extent also the concerns of the political elite, facilitated a type of enclavism that is the topic of this book. PSCs began to flourished. The insights from this section highlights how biopolitical abandonment post-apartheid is not unprecedented. It is the continuation of earlier processes, though executed through different means. The similarities between the beneficiaries and losers of apartheid era abandonment, and its form of overtly oppressive abandonment, and post-apartheid abandonment are however significant. The composition of the middleclass has changed, but the majority of the population still live precarious lives. Frontier governance dictates that groups are often divided through security infrastructures, which insulate many within the political and economic elites from the open frontier. These security infrastructures are associated with liberal logics.

Security infrastructures, frontier governance and the contradiction of closure through abandonment

Security infrastructures aid what Kempa and Sing (2008) call an individualised political peace as opposed to a social and economic peace. Private security entrenches the underlying assumptions around a belief in rights to and control over private property with minimum state interference. The individual is valued as the primary focus of rights and responsibilities, while economic growth remains the primary objective of any economic society (Kempa and Singh, 2008:341). Hence, Kempa and Singh (2008:342) argue that private security has played a key role in perpetuating the idea that the status quo is sustainable and desirable despite much evidence to the contrary.

Consequently, frontier governance has spawned a range of security infrastructures. From the 1970s PSCs have gradually become a common feature in South Africa. One company operating in Potchefstroom for example was formed the day after the 1976 Soweto uprising. Throughout the 1980s the police and military were focussed on the emergency that was dealing with insurgent liberation movements. As mentioned previously, PSCs were encouraged to look after national key points. After democracy's inception the industry grew exponentially. Today it employs more than double the staff of the SAPS (cf. Van Riet, 2020:82) and by 2007 South Africa had the largest private security sector in the world, as a component of gross domestic product (Abrahamsen and Williams, 2007:243).

Diverse security infrastructures have developed and grown over the years. The borders of the frontier are marked by the likes of high walls, razor wire, electric fences, alarm systems, laser beams and CCTV cameras on private and public land. These technologies divide relatively affluent households from the rest of the population, while these borders are patrolled by PSCs. Areas under the control of affluence expand by way of gated communities, commercial developments on private land, residents' associations and CIDs. The masses remain separated from the few in spatial terms, where open space, or in the case of Potchefstroom an industrial area, is used as a buffer zone.

As previously noted, there is a significant self-defeating consequence to the processes of closure through abandonment. This type of 'closure', taken to its logical conclusion, entrenches grave inequalities and widespread abject suffering, not to mention perceptions of superiority and of a perennial threat inherent to an Other that results from logics of equivalence. Often the relationship to this Other becomes reified through a set of security infrastructures that can be purchased (Van Riet, 2020:94). Consequently, class- and race-based conflict remains commonplace. But, the threat of the poor, unemployed black person to wealth and well-being, actual and perceived, cannot be alleviated. Abandonment by definition means condemning the masses to a lack of income and other opportunities. The polity functions for a small minority only and dysfunctions for a vast majority. The minority can never rest assured and require more and more security infrastructures to stave off often desperate or angry criminals. The disenfranchised masses represent an internal politics always potentially detrimental to the interests of the minority and they remain an available political resource for the opportunistic actions of populist movements, perhaps even increasingly so. Therefore, closure through abandonment and frontier governance, in the long run, cannot be in anyone's best interest. More inclusive strategies are required that eventually consolidate a shared sense of legitimacy. As mentioned, this will not require closing a frontier as much as it will require the complete abandonment of the frontier imaginary and governance and the gradual dismantling of its physical manifestations. Throughout this book I will highlight potentialities already present in the everyday politics of crime that might be leveraged and developed further to this end. Of course, the analysis in this book must be viewed in addition to macro-level political economy. The need to include the masses in the productive economy remains imperative. The following section further substantiates the claim that closure through abandonment is self-defeating, in particular, by assessing the nature and extent of crime in South Africa. Although marginality stemming largely from an abstract notion of abandonment cannot comprehensively account for crime, it does go some way towards doing so. The following section is intended to provide greater clarity regarding why and how abandonment and crime are related.

Explaining crime in South Africa

The National Crime Prevention Strategy (NCPS) of 1996 outlined numerous insights reflective of international thought on crime. It explained the roots of the 'current crime situation' in 1996 in the following manner. Periods of political transition have internationally been associated with increased crime. Illegitimate mechanisms of social control have been broken down, without immediately being replaced by others. In South Africa this is exacerbated by the disarray of other sources of social authority such as schools, traditional communities and the public service. To clarify this point, I can add that the migrant labour system discussed earlier and the need for women to be live-in housekeepers, raising white children, came to the detriment of their own families. Many children were raised in rural areas by their grandparents. Others were orphaned very young, often due to HIV/AIDS and subjected to an insecure childhood, both physically and ontologically. Rape for example continues to be committed against children, while children have been exposed to much uncertainty about their immediate future from a very young age. The transition to democracy created significant expectations that could not be met immediately, but, these limitations were dealt a devastating set of blows through austerity and the large-scale looting of potential sources of social spending. The NCPS noted that expectations upon the end of apartheid contributed to justifications for crime. Today, these senses of enduring injustice and the prominence of a disaffected youth might help explain many instances of crime.

Poverty in an absolute sense interacts with other socio-political and cultural factors to contribute to crime and the development of crime syndicates and youth gangs. But poverty is a multidimensional concept. According to the Oxford Poverty and Human Development Initiative (2020),

> Multidimensional poverty encompasses the various deprivations experienced by poor people in their daily lives – such as poor health, lack of education, inadequate living standards, disempowerment, poor quality of work, the threat of violence, and living in areas that are environmentally hazardous, among others.

According to Prinsloo (2003) multidimensional poverty is a major factor in marginalisation. It creates 'a culture of being vulnerable, powerless, isolated and physically and spiritually inadequate'. People often lack the basic necessities such as food, water and safety and opportunities to fulfil their potential. Multidimensional poverty involves fragmented care in female and child headed households, uninvolved parents, poorly equipped schools, demotivated and unqualified teachers and consequently indifferent learners. Poorly equipped schools and demotivated teachers are also manifestations of inequality in South Africa. Some children attend extremely well-funded private and public schools. School governing bodies are allowed to set school fees. If the learner catchment area is wealthy, then fees can be raised. The converse is also true. Regular school absenteeism and high drop-out rates are very common. To this we may add, based on my preliminary research in the neighbouring Ikageng police precinct, how membership of a gang offers youths a sense of belonging, as well as access to drugs. The latter sometimes fuels a habit that requires stealing. Being a member of a gang also places expectations of criminal activity on the member in order to belong. This might imply theft, but also violence.

The NCPS therefore espoused a developmental approach to crime prevention, set in a broader social context. It was reinforced by the 1998 White Paper on Security (South Africa, 1998), which emphasised the role of local government in crime prevention. The implication according to Newham (2005:1) was that crime prevention includes addressing socio-economic and environmental factors, as crime is not purely a security issue or as Lamb (2018:938) refers to it as espousing a more holistic approach to crime (also see Govender, 2018). Given the biopolitics of abandonment discussed earlier, it is no surprise that not much came of this approach to crime reduction. Crime continued to flourish, as there was no appropriate local level politics in place by which to implement the seemingly logical assessment of the NCPS.[10] Pelser (2008) furthermore notes that multi-agency crime prevention partnerships under the NCPS often became ends in themselves, without meaningfully contributing to their actual purpose. In addition, Lamb (2018:938) notes that the institutional culture within the SAPS made it an organisation inherently opposed to the type of cooperation required by the NCPS. We should also note the lack of resources and staff that station commanders have consistently had to deal with, along with wide spread allegations of corruption (cf. Burger and Grobler, 2017). It is not inconceivable that rank and file police officers were often drawn into the local government politics of patronage or drawn into associations with criminals, in part because of the precedent set all around them. Finally, various government departments were struggling to adapt to the multiple demands of the NCPS (Lamb, 2018:939).

By 2000 greater pressure from the populace to more effectively combat crime forced a change. Official vernacular changed to a 'war on crime', following the adoption of the National Crime Combatting Strategy (NCCS, cf. PMG, 2001). This document had a clearer focus on serious and violent crime (Lamb, 2018:938–939). It was also more in line with the history of the SAP/SAPS and the former freedom fighters that joined it after democratisation. Crime prevention was now equated to policing. Newham (2005), Govender (2018) and myself find ourselves more inclined towards the initial approach. In 2005 already Newham noted severely overcrowded prisons that continue to limit the possibility of rehabilitation. Convicts are typically released 'alienated, unable to find employment, and with better ties to criminal networks' (Newham, 2005:8). Many of the root causes of crime continue to exist. These causes include severe inequality, as is evidenced in South Africa's gini-coefficient of 0.67 in 2015 (StatsSA, 2019:5). Other causes linked to multidimensional poverty are high unemployment rates and absolute poverty, largely caused by unemployment and inequitable access to opportunities for upward social mobility. Adequate schooling is one example of the latter.[11]

In 2010 the police became even more militarised. It adopted military ranks and salutes. Crime was now to be kept in check through clinical operations in identified hotspot precincts (Lamb, 2018:940). This has often translated into the use of excessive force as the under-resourced SAPS became overwhelmed by everyday tasks (Lamb, 2018:942). Both Potchefstroom and Ikageng were included in the list of priority precincts. This has not translated into adequate staff for everyday policing. The focus moved from crime prevention to law enforcement. However, real crime prevention requires a set of longer-term activities. It requires contributions from various departments, such as social services, local economic development and I believe its effects are dramatically less in lieu of a meaningful, functioning macroeconomic strategy, which creates jobs in spite of globalisation. As Newham (2005:11), however, alludes, it is difficult to coordinate programmes across departments when these departments struggle to fulfil their core mandates. Again, given cadre deployment and the often problematic and stifling politics in local government and its adverse effects on a functioning bureaucracy, crime prevention was always going to be difficult. It should also be noted that with limited personnel and other resources and internal corruption, overfull court rolls and prisons, the ability of the SAPS to actually conduct law enforcement remains questionable. Statistics South Africa (2017:46) notes that districts with a high number of police stations and a low number of households per station have lower crime rates. This supports the logic that limited police resources can affect crime rates even after a lot of the funding for crime prevention was diverted to the criminal justice system. In my research in Potchefstroom and for that matter in Ikageng, I have found it very hard to see a 'war on crime'. Instead, what I have seen is a police service struggling to get by and a service that is increasingly bypassed by citizens who draw on private security and vigilante groups. This will continue and likely grow as long as the aforementioned root causes are not addressed.

I should now link what has been discussed in this chapter to theories of crime in order to make explicit the link between biopolitical abandonment and crime. It is far too simplistic to link crime to poverty only. This would not explain relatively low levels of crime in territories with generally high levels of absolute poverty. At the same time high levels of multidimensional and absolute (economic) poverty in South Africa are not irrelevant. According to Statistics South Africa (2017:34), criminals are more attracted to households that have more valuables than poor households. This observation offers the missing piece of the puzzle, linking multidimensional poverty to certain forms of crime through inequality.

Thus, I believe the insights produced earlier requires theorising informed primarily by the established tradition in Criminology and the Sociology of crime of Strain Theory.

Abandonment does many things. It reinforces inequality in material terms. It also does this while conspicuous consumption by both the political and economic elites are plain to see. So now we find abandonment in a territory that has a long history of ill-gotten gains. In this context senses of injustice may perpetuate a type of norm confusion, linked to multidimensional poverty. There is a lack of opportunities, such as access to education, low skilled work, health and even personal security in the form of parents that are present and able to provide for children. In such a context the tradition of Strain Theory, following from Merton's seminal paper of 1938, posits that people will revert to illegitimate means to obtain highly valued goals, when legitimate means are largely inaccessible. We may extend these goals very broadly from meeting basic needs to obtaining popular luxuries, but space does not allow for a comprehensive analysis. I think it is adequate and still instructive to reduce the use of Strain Theory to the absence of sufficient legitimate means for upward social mobility subsequent to a long and enduring history of injustice, characterised by various forms of inequality. In simple terms, absolute and multidimensional poverty matters in explaining crime in South Africa, but it is especially significant because of the critical mass of citizens who are relatively wealthy, for what are often perceived as dubious reasons. Biopolitical abandonment fuels the underlying material conditions referred to here and likely also perceptions of injustice. It is for these reasons that the quotation by Dieltiens (2011:96) cited in the introduction to this chapter, namely that 'punishment of crime retains its political and economic utility' makes most sense. Punishment of crime, in the absence of additional prevention strategies and remedial structures, perpetuates abandonment. Apart from the ample ethical concerns we may raise, this strategy is also practically unsustainable.

Conclusion

This chapter has sought to theorise the role of the security infrastructures in contemporary South Africa. This has been done by situating these infrastructures in a metaphorical reopened frontier that has developed alongside the unravelling of apartheid. Security infrastructures are part and parcel of attempts to close a perceived frontier through biopolitical abandonment. In its contemporary form abandonment is performed in the everyday through diverse actors above and beyond the state. The political economic logic of closure through abandonment, augmented by the spatial logic of frontier governance, is, however, self-contradictory, as it continues to fuel the root causes of crime. Frontier governance, as border-making is fused to notions of 'us' and 'them' and a certain finality or telos whereby hegemony, as an inherently overdetermined and contingent reality, can be overcome. The chapter has explained crime in South Africa and its link to biopolitical abandonment through the combination of absolute economic and multidimensional poverty and Strain Theory. We now need a conceptual understanding of the collective of security infrastructures that sustain enclavism. This is the topic of the following chapter, where the concept of the *laager* and its relevance to crime and security is explained.

Notes

1 This figure is based on conversion rates on 22 February 2021, at around ZAR 15 to USD1.
2 The BEE programme was initiated under President Mbeki in the 1990s. Its overt objective was to grow the black middleclass. One major way in which this took place was through state tenders. The process at times led to significant concentration of wealth, not unlike large white controlled conglomerates. This has left questions as to how broad based this initiative was in actual fact.
3 The term *Afrikaner* is clearly derived from the word Africa. It is a noun that refers to a people, typically based on race and language. They are white settlers in South Africa and some neighbouring countries, who speak Afrikaans, a language mostly derived from Dutch. Most people who view themselves or who are viewed by others as Afrikaners are of a combination of Dutch, French, German and British decent.
4 *Boer* is the Afrikaans word for farmer. However, as a political concept it holds at least two distinct connotations. It is either spoken with relative pride by Afrikaners to refer to themselves or it is used as a racial or ethnic slur spoken by black or English-speaking white South Africans against this group. I use the term here because the literature typically refers to 'Boer republics' and not 'Afrikaner republics'. If it is not clear yet, please note that the term in these instances refers to those who governed these territories and not to the general population who resided in those territories.
5 The ZAR and Free State Republics later become two of the four provinces of the Union (under the British flag) and later Republic of South Africa. The name of the ZAR changed to the Transvaal (across the Vaal River). With democratisation in 1994 the country was demarcated into nine provinces. With this demarcation the Free State was one of only two former provinces (the other being Kwazulu-Natal, previously known only as Natal) that remained.
6 The term South African War (1899–1902) has become the more appropriate name for what was previously known as the Anglo-Boer War. This is because black South Africans were also involved in the war in various capacities.
7 Bantu Education was the system of education designated for Black South Africans. It was inferior to the curriculum taught to white children, as the argument was made that different races generally had different innate levels of potential.
8 The AWB is a far-right white supremacist and separatist paramilitary organisation formed in 1973. Their headquarters is in the small town of Ventersdorp, about 50 km from Potchefstroom, which is also part of the JB Marks Municipality.
9 The 'Battle of Bophuthatswana', was an event early 1994, where the AWB entered the 'independent homeland' of Bophuthatswana supposedly to keep its leader Lucas Mangope in power. Mangope's reign was under pressure, as the movement towards democratisation in South Africa, including the dismantling of 'ethnic' homelands reached a critical mass.
10 For more insight into this distinction between progressive policy and an appropriate politics by which it can be implemented see Pithouse's (2009) analysis of the Breaking New Ground housing policy and its implementation.
11 The Gini-coefficient is a widely recognised measure of income inequality within a territory. A Gini-coefficient of 0 represents absolute equality and 1 represents absolute inequality, StatsSA (2019:5) note that South Africa, for as long as there has been reliable national level data, has consistently experienced extreme income inequality, which has conservatively placed it in the top five most unequal territories in the world, at least.

References

Abrahamsen, R. and Williams, M.C. 2007. Securing the city: Private security companies and non-state authority in global governance. *International Relations*, 21(2), pp. 237–253.

Agamben, G. 1995. *Homo Sacer: Sovereignty and bare life*. Stanford: Stanford University Press.

Arendt, H. 1953. Ideology and terror: A novel form of government. *The Review of Politics*, 15(3), pp. 303–327.

Bond, P. 2005. *Elite transition: From apartheid to neoliberalism in South Africa*. 2nd ed. Pietermaritzburg: University of Kwazulu-Natal Press.

Buhlungu, S. 2009. The rise and decline of the democratic organizational culture in the South African labor movement, 1973 to 2000. *Labor Studies*, 34(1), pp. 91–111.

Burger, J. and Grobler, S. 2017. Why the SAPS needs an internal anti-corruption unit. Policy brief. Institute for Security Studies. Pretoria. Online: https://issafrica.org/research/policy-brief/why-the-saps-needs-an-internal-anti-corruption-unit. Date of access: 16 March 2020.

Desai, A. 2018. The Zuma moment: Between tender-based capitalists and radical economic transformation. *Journal of Contemporary African Studies*, 36(4), pp. 499–513.

Dieltiens, N. 2011. The making of the criminal subject in democratic South Africa. Master's thesis. University of the Witwatersrand. Braamfontein.

Ellis, S. 1998. The historical significance of South Africa's third force. *Journal of Southern African Studies*, 24(2), pp. 261–299.

Foucault, M. 2008. *The birth of biopolitics: Lectures at the College du France 1978-1979*. Translated by Graham Burchell. New York: Palgrave Macmillan.

Franks, P.E. 2014. The crisis of the South African public service. *Journal of the Helen Suzman Foundation*, 74, pp. 48–56.

Govender, D. 2018. Contact and property related crimes in South Africa: Need for strategies and democratic policing. *International Journal of Criminal Justice Sciences*, 13(1), pp. 55–67.

Kempa, M and Singh, A. 2008. Private security, political economy and policing of race. *Theoretical Criminology*, 12(3), pp. 333–354.

Koelble, T.A. 2018. Globalization and governmentality in the post-colony: South Africa under Jacob Zuma. Working paper. WZB Discussion Paper, No. SP V 2018–103. WZB Berlin Social Science Center. Online: www.econstor.eu/handle/10419/190803. Date of access: 16 March 2020.

Lamb, G. 2018. Police militarisation and the 'War on Crime' in South Africa. *Journal of Southern African Studies*, 44(5), pp. 933–949.

Leggasick, M. 1980. The frontier tradition in South African historiography, in S. Marks and A. Atmore (eds.) *Economy and society in pre-industrial South Africa*. London: Longman, pp. 44–79.

Lemke, T. 2011. *Biopolitics: An advanced introduction*. New York: New York University Press.

Liu, A. and Saal, D.S. 1999. An input output analysis of structural change in apartheid era South Africa: 1975–93. Working paper. Online: www.iioa.org/conferences/13th/files/Saal_AparthiedChange.pdf. Date of access: 14 March 2020.

Marais, H. 1998. *South Africa: Limits to change: The political economy of transition*. London: Zed Books.

Merton, R.K. 1938. Social structure and anomie. *American Sociological Review*, 3, pp. 672–682.

National Income Dynamics Study – Coronavirus Rapid Mobile Survey (NIDS/CRAM). 2020. Overview and findings NIDS-CRAM synthesis report wave 1. Online: https://cramsurvey.org/wp-content/uploads/2020/07/Spaull-et-al.-NIDS-CRAM-Wave-1-Synthesis-Report-Overview-and-Findings-1.pdf. Date of access: 22 July 2020.

Newham, G. 2005. A decade of crime prevention in South Africa: From national strategy to local challenge. Research report. Centre for the Study of Violence and Reconciliation. Cape Town. Online: www.csvr.org.za/docs/policing/decadeofcrimeprevention.pdf. Date of access: 16 March 2020.

Oxford Poverty and Human Development Initiative (OPHDI). 2020. Policy: A multidimensional approach. Online: https://ophi.org.uk/policy/multidimensional-poverty-index/ Date of access: 22 July 2020.

Parliamentary Monitoring Group of South Africa (PMG). 2001. SAPS transformation; National Crime Combatting Strategy: Briefing. Online: https://pmg.org.za/committee-meeting/553/ Date of access: 6 November 2020.

Pelser, E. 2008. Learning to be lost: Youth crime in South Africa. Discussion paper for the HSRC youth policy initiative, Reserve Bank, Pretoria, 13 May. Online: www.cjcp.org.za/uploads/2/7/8/4/27845461/hsrc_youth_crime_discussion_paper2.pdf. Date of access: 16 March 2020.

Pithouse, R. 2009. A progressive policy without progressive politics: Lessons from the failure to implement 'Breaking New Ground'. *Journal of Town Planning*, 54, pp. 1–14.

Povinelli, E. 2011. *Economies of abandonment: Social belonging and endurance in late liberalism*. Durham, NC: Duke University Press.

Prinsloo, E. 2003. At society's margins: Focus on the youth in South Africa. *Educare*, 32(1–2), pp. 275–292.

Rasool, C. 2006. History anchored in politics: An interview with Martin Legassick. *South African Historical Journal*, 56(1), pp. 19–42.

Ross, R. 1981. Capitalism, expansion, and incorporation on the southern African frontier, in H. Lamar and L.M. Thompson (eds.) *The frontier in history: North America and South Africa compared*. New Haven: Yale University Press. pp. 209–233.

Schneider, G.E. 2000. The development of the manufacturing sector in South Africa. *Journal of Economic Issues*, 34(2), pp. 413–424.

South Africa. 1998. White paper on safety and security. Online: www.policesecretariat.gov.za/downloads/white_paper_security.pdf. Date of access: 6 November 2020.

South Africa. 1996. National crime prevention strategy. Online: www.gov.za/documents/national-crime-prevention-strategy-summary?gclid=EAIaIQobChMI49vQgYju7AIVibHtCh3vrwtCEAAYASAAEgLHRvD_BwE. Date of Access: 6 November 2020.

Southall, R. 2009. Understanding the 'Zuma Tsunami'. *Review of African Political Economy*, 121, pp. 317–333.

Statistics South Africa (StatsSA). 2020. Unemployment rises slightly in third quarter of 2019. Online: www.statssa.gov.za/?p=12689. Date of access: 13 March 2020.

Statistics South Africa (StatsSA). 2019. Inequality trends in South Africa: A multidimensional diagnostic of inequality. Report No. 03–10–19. Online: www.statssa.gov.za/publications/Report-03-10-19/Report-03-10-192017.pdf. Date of access: 22 July 2020.

Statistics South Africa (StatsSA). 2017. Exploring the extent of and circumstances surrounding housebreaking/burglary and home robbery: An in-depth analysis of the Victims of Crime Survey 2015–2016. Report number 03–40–04. Pretoria: Statistics South Africa.

Swilling, M. 2019. Can economic policy escape state capture? *New Agenda*, 72, pp. 24–27.

Terreblanche, S. 2002. *The history of poverty and inequality in South Africa*. Pietermaritzburg: University of Kwazulu-Natal Press.

Thebe, T.P. 2017. Political education for good governance in South Africa's local government and communities. *Journal of Public Administration*, 9(5), pp. 123–135.

Van Riet, G. 2020. Intermediating between conflict and security: Private security companies as infrastructures of security in post-apartheid South Africa. *Politikon: The South African Journal of Political Studies*, 47(1), pp. 81–98.

Van Riet, G. 2016. The limits of political development and constitutionalism in South Africa. *New Contree*, 75, pp. 98–115.

Van Riet, G. 2014. Instrumental reason and neoliberal governmentality: A critical analysis of disaster risk assessment and management in South Africa. DLitt et Phil thesis. University of Johannesburg. Aucklandpark.

Visser, W. 2004. Trends in South African historiography and the present state of historical research. Paper resented at the Nordic Africa Institute. Uppsala. Online: www.academia.edu/2874752/trends_in_south_african_Historiography_and_the_present_state_of_Historical_research. Date of access: 16 March 2020.

Walker, E. 1928. *A history of South Africa*. London: Longman, Greens & Company.

Wolpe, H. 1972. Capitalism and cheap labour-power in South Africa: From segregation to apartheid. *Economy and Society*, 1(4), pp. 425–456.

2 The laager as a collective security infrastructure

Introduction

The previous chapter conceptualised a contemporary reopened frontier in South Africa. This metaphorical frontier, faced with the contradictions of closure through biopolitical abandonment, has created a need for the political and economic elites to protect themselves from crime through frontier governance. For some, crime is linked to a ubiquitous and largely equivalent Other. For others, protection from crime is, at least mostly, just about safety. Still, the country has increasingly become the site of enclaves. The current chapter continues where the previous ended, by conceptualising the very enclaves that have come to characterise the South African landscape, including Potchefstroom. These enclaves include gated communities, CIDs and privately policed suburbs outside gated communities.

The laager as a historical phenomenon may assist us in understanding contemporary enclaves within the frontier. The logic by which public and especially private crime fighting resources are deployed, and the materiality of their deployment, do not exist in isolation. Many premises, intellectual frameworks and physical resources drawn upon in the response to crime are taken for granted. An infrastructural approach to social research helps to lay bare underlying structures for their role in allowing and constraining the circulation of ideas, practices and people. This chapter is particularly concerned with a non-continuous physical and metaphysical laager that shapes many of the practices of crime fighting in urban Potchefstroom. This contemporary laager might be defined as the sum total of physical, virtual and ontological structures that facilitate border-making and enclavism in contemporary societies, where there is a symbiotic relationship between the physical and the ontological. The laager, just like its initial manifestation in the 1800s, as a circle of wagons, also represents fear. Today it is a defensive position against crime, but also much more. This chapter will only introduce this understanding of enclavism, akin to what Lemanski (2006) calls 'fear of crime plus'. What exactly the 'plus' entails will be revealed throughout this book. In this chapter I will also briefly consider the relevance of the laager for other contexts. My argument is that the laager is potentially of use as a lens through which to understand responses to crime beyond the work of the state police. Although the word laager is South African, the laager as a historical

DOI: 10.4324/9781003028185-4

phenomenon is not unique to South Africa. Moreover, as is explained in the following, the laager might, based on the needs of a particular analysis, be more useful than related terms such as 'semigration' (cf. Ballard, 2004) and fortress (Blakely and Snyder, 1997).

The following section of this chapter explains the significance of viewing the laager as infrastructure and as a central or pivotal infrastructure in the context of an infrastructural approach to crime and security. The laager may be viewed as a nodal infrastructure, derived from Laclau's notion of a nodal or central signifier. However, because this might cause confusion with Shearing's (cf. Shearing and Wood, 2003) oeuvre focussed on nodal governance. I have, therefore, chosen to frame the laager as a pivotal infrastructure in the politics of crime and security. This framing does, however, follow from a logic by which Laclau is brought into conversation with an infrastructural approach to crime and security. This approach is contrasted to commonly held notions of the laager as a mentality only. Thereafter, the chapter explains how the laager is enacted through security discourse and practice in contemporary Potchefstroom. The analysis then offers an initial answer to the following question: What do security infrastructures do in contemporary Potchefstroom? In other words, I seek to explain how infrastructure as a substrate for the circulation of ideas and practices often manifests unfavourable social practices. This section also explains how crime and other subversive practices trouble the laager. Crime, in particular, is not a constructive way of troubling the laager. Consequently, the chapter then considers more constructive ways to unsettle the laager, based on the potential inherent in some existing social practices, and it concludes by similarly considering the relevance of the laager for analyses elsewhere. Thus, in addition to elaborating the laager as infrastructure, the chapter is critical by giving a preliminary overview of the politics of crime and security in Potchefstroom. More focussed chapters that follow in Part II and Part III of this book will elaborate themes from the current chapter, before we once more return to more general conclusions in the final chapter.

The laager as infrastructure

Infrastructures that deal with crime prevention and response are diverse. These infrastructures include the SAPS, PSCs, social media platforms, neighbourhood watches, CPFs, the news media, homeowners' associations, alarm systems and electric fences. All of these infrastructures may be related to the broader concept of crime prevention through environmental design, which includes elements such as a positive image that facilitates local ownership of the area in question, opportunities for surveillance and clearly defined boundaries. Environmental design also includes target hardening, that it, making a potential target less easily penetrable (cf. Cozens et al., 2005:330). The laager presents a central scaffold around which other infrastructures and practices are shaped. It is in a sense the sum total of all physical, virtual (such as social media platforms) and ontological security structures as they manifest in South Africa, often through the exclusionary practices of frontier governance. A laager, according to the *Oxford Living*

Dictionary (online), is defined as 'an encampment formed by a circle of wagons' or 'an entrenched position or viewpoint that is defended against opponents'. The former definition attests to an era in South African history where the trekboere mentioned in the previous chapter, moving inland from the Cape Colony, would create such a circle of ox wagons to establish a defensive position from which they could ward off potential enemies.

The use of a laager was and is not unique to South Africa. In the USA settlers moving West from the East employed a similar structure, primarily to keep their livestock in, and occasionally, in fear of less frequently realised attacks by Native Americans. Moreover, globally, the phrase 'to circle the wagons' is a common figure of speech that now refers to much more than the initial practice. Consider the following definitions of the idiom, 'to circle the wagons'.

> 'Circle the wagons' or 'circle your wagons' means that members of a team or group must work together to protect themselves from some outside danger. Generally, they prepare for a possible attack. Some language experts say this expression comes from the time when many Americans were moving away from the East Coast to what is now the western United States.
>
> (Learning English, 2017)

> Look for protection, get defensive, get ready for an attack; from the old west where the pioneers would circle their wagons for protection from the Indians.
>
> (UrbanDictionary.com)

> If you circle the wagons, you stop communicating with people who don't think the same way as you to avoid their ideas. It can also mean to bring everyone together to defend a group against an attack.
>
> (UsingEnglish.com)

These definitions cover the scope of the idiom. It can relate to defensiveness, insularity, avoidance of some groups or people and to frontiers. Especially the last definition invokes notions of what Laclau and Mouffe (2014[1985]:117) termed logics of equivalence. Both the notions of an inside group and of an outside group are sutured together, despite the diversity within both groups. Moreover, definitions of inside and outside are co-constitutive. The one cannot exist without the other, although the members of the inside group who do the definitional work are typically ignorant of this unmistakable relationship between inside and outside. Moreover, based on these definitions, it should be clear that the application of the term extends far beyond South Africa. The origin of the idiom 'to circle the wagons' is not South African. In South Africa, the word laager however emerged in reference to manifestations of the aforementioned practices.

Conservative South Africans have often been accused of displaying a 'laager mentality' (Harris, 2002; Nauright, 1996). Especially white and Afrikaans middleclass citizens have been known to revert to newer incarnations of this structure. Accusations of forming a metaphorical laager post-apartheid have typically been the result of moves to preserve language and culture through insular practices, such as the management of admissions by school governing bodies and restricting access to public spaces through privatisation (Soudien and Sayed, 2004; Harris, 2002). My argument, however, is that while the notion of a laager mentality has much validity in reference to narrowly defined nationalist and class-based aspirations, there is significant potential in viewing the laager as the collective of physical and ontological infrastructures that allows some ideas, practices and people to circulate while restricting the circulation of others. Potchefstroom is a useful case study in this regard. It is a space where hegemonic political, economic and cultural practices remain largely middle class, white and Afrikaans. But, for the most part, it does not explicitly exclude people based on language and race per se. Many within the political elite also choose to settle within the laager for reasons of security and status. Therefore, if one is to speak to the South African literature on the laager, we must acknowledge that the laager is no longer exclusively a narrowly conceived racial or ethnically mobilised devise. The realities of crime and the vagaries of relative status are changing this historical association. Having said that, Potchefstroom, is arguably somewhat behind the rest of the country in terms of racial integration. Of this there is much evidence in this book. Many of those of all races who can afford to live inside the laager do so. The various infrastructures listed here have largely been formed to serve their interests and continue to do so. What is at stake in my interpretation of the laager compared to others, is the difference between a mentality that is more or less general and bounded and more or less always an independent variable, and a perspective from which the laager is part of a dynamic and rich tapestry of material and ontological structures, processes and ideas that shape a more layered and textured reality. The book attempts to excavate the richness and diversity of this reality as points of interference and hope for progressive praxes. Stated differently, we might say that some of the insights one would gather from the 'laager as mentality' approach remain. However, the 'laager as infrastructure' approach, in addition, allows us to recognise more of the nuances of daily life and the opportunities for constructive intervention, based on the flows allowed and constrained by the various infrastructures by which the laager is shaped. This mode of thinking also accounts for how the laager is shaped and how it is dynamic, just like the rest of social reality.

There is one final way in which the chapter contributes to the existing literature. There has been much research on gated communities in recent decades (Jürgens and Gnad, 2002; Low, 2001; Blakely and Snyder, 1997). Although gated communities do exist, they are not as common in Potchefstroom as they are in the large cities, such as Johannesburg and Cape Town, which are far more frequently studied. Residents of Potchefstroom, generally, do not live in gated communities. Therefore, Potchefstroom allows us to observe the laager at work, shaping or being shaped by dynamic interactions with other infrastructures, ideas and practices, outside of gated communities. In addition to extending the field of empirical application associated with the laager and middleclass insularity, this perspective is significant as the analysis challenges the laager's association with such insularity purely within communities surrounded by a single wall. Some might still argue that metaphors such as 'semigration' (Ballard, 2004) or 'fortress' (Blakely and Snyder, 1997) are preferred to the notion of a laager. My response would be that choice of concept will vary depending on the context of the analysis. It is, as such, an empirical question.

Ballard coined the term 'semigration' to refer to people who choose to live in gated communities, often designed based on European architectural aesthetics. Of this Tuscan influenced architecture is arguably the most well-known example. They might not be able to emigrate for a number of reasons. They may choose not to acknowledge consistently their South African and African reality. The term 'fortress' was used to explain the growth in gated communities in the USA since the latter quarter of the 20th century. Here the emphasis seems to be somewhat more on escaping crime, although the universal effectiveness of gated communities as a strategy to evade crime has been disputed (cf. Lang and Danielsen, 1997). The merits of drawing on the concept of the laager, as defined earlier, in an infrastructural approach to security, is in its status as the sum total of physical, virtual and ontological structures that manifest to facilitate diverse forms of enclavism. These go beyond purely gated communities and they go beyond merely physical structures. They are rooted in everyday flows of ideas, practices and people facilitated and constrained by an extensive infrastructural substrate.

The discussion now turns to how the laager is enacted and reiterated through the symbiotic relationship between material and ideational factors. At the same time the existence of the laager and the imaginary it inflects into the social, facilitates the circulation of problematic ideas and practices.

Enacting and reiterating the laager

Comments on social media give much insight into the movements of its users. The reality is movement between enclaves. Motor vehicles are an important infrastructure facilitating this movement. They serve as mobile pockets of relative security between more formally guarded and governed private spaces. The laager ensures that there are frequently attempts to securitise these vehicles and a diverse set of other objects and geographies. Securitisation will be dealt with in greater depth in Chapter 6. Here, it can be described briefly as, following the Copenhagen School of security studies, rendering an object as subject to a threat and therefore in need of protection. Securitisation may also imply that a particular agent is a threat in need to be dealt with through security practices (compare Buzan et al., 1998). This book is primarily concerned with routine practices and as such more in tune with PARIS School's interpretation of securitisation as the establishment of routine security practices and forms of expert knowledge (see Bigo and McCluskey, 2018). There are, however, frequent attempts in Potchefstroom to render a security concern acute and in need of extraordinary measures.

The aforementioned fortified elements range from private homes to neighbourhoods, suburbs and privately owned commercial areas such as malls, to very public spaces and formal infrastructure. By 'formal infrastructure' I mean the more traditional narrow conception of the term, including roads, railways, and water and electricity infrastructure. The laager as a manifestation of frontier governance is a barrier beyond which middleclass and often-ethnic interests can supposedly flourish. Such flourishing is, however, perennially endangered by the internal contradictions of closure through abandonment. Infrastructures aid attempts by some to '(re)colonise' public spaces in a reopened frontier. In a sense, this means that a minority of residents want to colonise formal infrastructure, and the laager is the intersubjective substrate and vehicle that facilitates the circulation of such ideas and practices. This intersubjectivity is aided by the very real set of physical structures, some remnants of apartheid, that already direct and allow for the direction of flows. In Potchefstroom the 'buffer' that is the industrial area between Potchefstroom and Ikageng is a key example.

The aforementioned ambitions are lubricated through an indirectly articulated discourse of onslaught from crime, among other aspects. Here I want to draw a parallel with official state discourse towards the end of apartheid, while distinguishing between the two. The total onslaught, coined by the minister of defence Magnus Malan, was accompanied by the invocation of a communist and insurgent threat to the white public and calls for increased militarism in the form of a 'total strategy'. This counter-insurgency strategy also included diplomatic and other political aspects (Botha, 2008:85). Since then the 'red peril' has been replaced by a 'black peril' of sorts. The insurgent is no longer a black communist (Posel, 1984). He is a black and often poor, typically male, criminal. Today, the calls for increased force are associated with hiring PSCs to patrol and secure the suburbs. These calls are also associated with fatalistic utterances associated with 'state failure' and generally gloomy futures. Crime is used as a marker of such decline. Comments such as 'We are under surge in Potch' (sic JB Marks Security Facebook group, 7 September 2017) are commonplace on social media. Another person suggested the use of a *sjambok* (a heavy leather whip) and police dog on vagrant intravenous drug users. Based on instances where I accompanied PSCs on patrol, I can state that such drug use happens especially in a particular stretch of sidewalk in front a building quite infamous for being largely taken over by criminal activities. It is not a part of town frequented by the middle class.

There is a common perception of a double standard in policing amongst social media users, whereby criminals are allowed to continue with criminality while law-abiding citizens, acting in response to crime are policed. There are also vague references to a subjectively defined group and their moral order being under attack. Once again, the precedent from the 1980s is clear. The moral order that is capitalist and Christian and often ethnically 'pure' has been reiterated, but as a less precisely sutured category.

Here, 'onslaught' is in a sense divorced from a broader whole, likely because that broader context is not always entirely visible on account of the materiality of the laager. Onslaught becomes the act of a political other, while the speaker professes to be an apolitical protagonist. However, as demonstrated in the previous chapter, crime is a problem of the broader social order, in no small way as a consequence of how different subject positions relate to each other. Subject positions as opposed to the notion of subjectivity are context bound performances. They are temporary and partial manifestations of the possible suite of positions taken by a person or entity, that are potentiated by contemporary discursive structures (see Laclau and Mouffe, 2014[1985]:102). In other words, these comments are blind to the realities of abandonment and the broader socio-political order that facilitates much of the crime the country faces. Instead, the implied course of action required may at times be conservative and (in)advertently in favour of closure through abandonment. This is furthermore significant in the case of vagrants sleeping in a field being portrayed as 'matter out of place' in the way Mary Douglas (1966:44) conceives of dirt. It is something or someone that drastically contravenes typical notions of purity and therefore it does not belong. Omitting the fundamentally political nature of 'outsider' production, allows for further colonisation of open spaces by the supposed protagonists. If 'outsiders' are matter out of place, as this logic suggests, then the situation can and should apparently be rectified, often by contacting PSCs to harass the persons in question.

I should make two important points here. Firstly, and to reiterate, although social media and other security infrastructures, such as the CPF are dominated by white and largely Afrikaans people, and by extension their interests, members of the political elite have also chosen to reside inside the laager. They also make use of security infrastructures, such as PSCs. Secondly, this critical discussion does not mean that crime is not real and that residents of Potchefstroom should not be vigilant in the face of relatively high levels of crime. Moreover, the onslaught conceived of today is different from that of the 1980s. Crime is far more real than the communist onslaught from the North purported by the NP government in the 1980s, used to justify extreme or even extraordinary measures. The discourse on crime however becomes problematic when other ideas, such as onslaught by a supposedly mysterious vindictive criminal and threats to cultural practices and the Afrikaans language are attached to it. The act of adding such ancillary discourses to the problem of crime, might be construed as a type of articulation aided by the reality of existing spatial forms. The very notion of matter out of place would have been impossible had the laager not been a metaphysical *and* material reality.

Exclusionary spaces of representation, in the Lefebvrian (1991) sense, have continued to hold sway over progressive representations of space. The partial changes in this regard brought on by democratisation have evoked many expressions of concern and anxieties. The following non-exhaustive examples demonstrate the point.

There have been various calls for mobilisation and solidarity in light of the apparent onslaught. Consider the following:

> Let's take hands and make a stand! Whether you were a victim of crime or you know someone who have (sic) been a victim of crime . . . Come and take hands with us in order to make a stand and say **enough is enough** . . . Crime

is devastating and **it's crippling our community!**' . . . 10h00 at 04 November 2017, Potch Dorp Rugby Club, 2 Piet Bosman STR, Potchefstroom. Please wear black to symbolise a community in mourning! If your business supports this initiative, please wrap a tree at your entrance in black. The historic Oak Avenue will also be wrapped in black.' [*sic*].

(JB Marks Security Facebook group, 25 October 2017)

Mobilisation against the siege from mysterious insurgents once more resonates with the 1980s jargon of a total onslaught. Today, just as in the 1980s, these citizens or the 'in-group', remains conceived of in a narrow sense. As was the case in the 1980s, the insurgent is devious. Consider the following comment: 'They are very quick and calculated' (JB Marks Security Facebook group, 29 June 2017).

At the same time, the notion of state failure and failures in municipal service delivery interact with fears of crime. Formal infrastructure remains at risk on account of poor municipal service delivery. Of course, there are instances of poor municipal service delivery. Including such failures into a sutured narrative of onslaught is, however, problematic. There is disillusionment with democratic government, the public service and shifting power relations. The SAPS too are linked to the onslaught, because they sometimes refuse to react to supposed threats in the way the public demands. To reiterate, the laager is an infrastructure that aids the representation of the remnants and reiterations of the old order as natural. 'Breaches' of the laager are viewed as distinctly unnatural and a reason for significant concern. In the democratic era, the police no longer serve these ends. In combination with limited resources and some instances of poor policing this means that the SAPS are considered by some as part of a failing state and, as such, as part of the enemy. Distrust in the police correlates with relative trust and praise for PSCs. Words of praise are often spoken for citizens' arrests and non-state actors who participate in crime fighting. The opposite often applies to the SAPS. Consider the following quotations:

> Thank you very much (name of security company) for your fast action at the dam area. You are the new police of our town. Potch SAPS, you are pathetic to say that you will not come out because it is a waste of time.
>
> (Author's translation, JB Marks Security Facebook group,
> 22 January 2017)

> Oh please, don't even try to ask the SAPS for help. They are beyond useless.
> (Author's translation, JB Marks Security Facebook group,
> 8 August 2017)

For many residents, the *volk* are under attack; so too are middleclass interests. As the earlier evidence suggests, this relates to a type of loosely sutured, though floating in terms of what it signifies, 'purity' linked to ethnicity, class, religion and/ or notions of respectability. Purity in this sense is associated with particular types of urban life, behind the physical barriers that form part of the laager, such as high walls, electric fences, the approximate borders of unfenced neighbour- hoods and suburbs. Behind high walls these forms of urban life employ animals, in particular dogs, as duel infrastructures. They are part and parcel of urban escap- ism inside the laager. Affection and pleasure circulate in part by drawing on these animals. At the same time dogs are security infrastructures, meant to warn off, warn of, and even attack, would-be intruders. The phenomenon of dog poison- ing therefore poses a double threat to certain modes of urban living, the order of things and life itself.

Life largely lived behind high walls, in shopping malls or other largely white and middleclass urban spaces, where motor vehicles for the most part ensure safe passage between these locales, may sometimes be linked to the notion of semigration (Ballard, 2004). It is a type of urban escapism from unwanted reali- ties. However, the story of security in urban Potchefstroom, in light of apparent criminality, is not one of semigration alone. Rather, laagerisation or manifest- ing the laager as an essential part of frontier governance, tends to vacillate between semigration and the (re)colonisation of public space. Often both are employed at the same time. As the metaphorical frontier remains open, and rearticulated, there are efforts to gain control over more space and to preserve control where it already exists. The use of PSCs, often summoned through pub- lic WhatsApp groups to clear the streets of people viewed as matter out of place, is one example in this regard. Another example is the call on social media for more security guards in public spaces. In general, comments on social media tend to securitise diverse objects, including the home, the neighbourhood, the car, public roads and streets. These attempts attest to the presence of a dynamic or non-continuous laager and an imaginary facilitated by an existing and reiter- ated socio-material structure. They are also attempt's at gradually extending the laager in ways most favourable to a small minority. Some of these practices of recolonisation are not necessarily wholly deliberate. Initiatives by home own- ers to set up CCTV cameras in the street or the initiative by the university and its partners to establish a CID are arguably primarily a response to crime. These structures do, however, represent in a *de facto* way instances where insu- lar spaces of representations have superseded official representations of space, based on notions of freedom of movement and desegregation (cf. South Africa, 1996). In this context recolonisation implies expanding the material reality of the laager, which also then informs and helps to the shape the politics whereby some members of the public and security professionals respond to crime. So, to summarise, the laager is enacted, as modes of border-making, urban escapism and as part of the self-contradictory and unsustainable techniques of closing the frontier through biopolitical abandonment and frontier governance. Laa- gerisation, manifests as semigration and the (re)colonisation of public spaces. These processes are instigated and allowed by the political and economic elites, motivated both by concerns over crime and problematic ancillary discourses, such as relative status and onslaught. The latter is mostly true for members of the economic elite.

Infrastructure in action

The aforementioned discussion explained how the laager is enacted and reiterated through the symbiotic interaction between the material and the metaphysical in shaping contemporary imaginaries. I will now elaborate further on how the laager interacts with other infrastructures and discourses in shaping reality. In other words, this brief section considers further the implications of the existence of a laager. Security infrastructures often help to securitise public spaces, setting in motion attempts to (re)colonise these spaces. Very often it is vagrants who are policed. Consider the following example:

> People in Grimbeeck Street opposite the Vodacom tower. There is bedding in the corner of the field next to my flat. I live in . . . I just want to warn everyone to listen and to be cautious in the area. Two housebreakings last night after 02:00 at . . . town houses, Hattingh Street, Bailie Park. Windows are removed neatly and put down. At one town house the people did not even wake up. That is how quiet they work.
>
> (Author's translation, JB Marks Security Facebook group,
> 20 June 2017)

Here the complainant infers a relationship between particular vagrants and instances of housebreaking. The police serving the public at large are not always amenable to intervening in such instances. The public then enlist private security to intervene often beyond their mandate.

> (We) phoned the police. The only reply we got was that 'they are homeless. They can be wherever they want to be.' (Name of private security company) removed them.
>
> (JB Marks Security Facebook group,
> 14 June 2017)

Such instances speak to the colonising of public space by the membership of a relatively privileged general 'in-group'. Residents summon security companies to particular scenes where fellow citizens are portrayed as matter out of place. They pose a threat to the securitised purity of public spaces. To be fair, we should acknowledge mitigating factors when asking why PSCs respond in these instances? The private security market is saturated. According to one PSC manager (Interview, June, 2018), there are 22 PSCs in Potchefstroom. Based on the 2011 Census, this implies one PSC for every 2879 people (StatsSA, 2017). It might therefore be difficult for security companies to refuse such demands by paying clients, as it could mean a loss of current clients and an unfavourable reputation among potential customers.

One may speculate that fear of crime is intertwined with fear of decline in property values. The fear of low-cost housing being built, particularly in the more affluent parts of Potchefstroom, appears to be viewed as a threat to complainants' wealth. Moreover, the implication is that the entire historically white part of the municipality, a material reality, that is Potchefstroom, should remain so, or it should at least be middle class and mainly white. Consider the following example.

> I see the newest informal settlement has been erected today opposite Duet-congregation in Mooirivier Avenue next to the canal. Nice and close to a water source and a stone's throw from the closest supermarket. Nice! So . . . how quickly is action required before it becomes formal housing?
>
> (Author's translation, JB Marks Security Facebook group,
> 9 March 2017)

Legislation may be part of the infrastructural substrate and work with other infra-structures to facilitate control over the flow of people in a democratic, though arguably largely unjust social order. The legislation in question includes those governing citizens' arrests, property rights and land occupation. Moreover, the limited capacity of the SAPS may work against ordinary people, as their right to freedom of movement is impeded by occasional harassment by other citizens and security companies.

Appeals to a threatened moral order assist the functioning of the laager. A legiti-mate case may of course be made for crime contravening reasonable notions of morality. Invoking morality can, however, aid division in society by providing justification for attempts to banish perpetrators to 'their own areas', and the colo-nisation of public spaces by the few. This is evidenced in the case where crime in a particular part of Potchefstroom has set in motion plans for a CID. Without pre-empting the discussion in Chapter 9, dedicated to this CID, I might impress that the literature on CIDs suggests that these infrastructures often provide room for actors such as PSCs to harass citizens and destroy or remove homeless peoples' belongings (cf. Paasche et al., 2014). The fact that there were discussions about fencing-off a park that is presently a public space does not bode well for the pros-pects of alternative approaches to managing such zones. This seems a rather drastic remedy, in conflict with freedom of movement and the right to the city as elabo-rated by Henri Lefebvre. Lefebvre's position is based on an aversion to systems of thought that simplify urban realities (Lefebvre, 2000:63). These systems paint over the divisions and inequalities in a society, or social order in Laclauian terms, by closing themselves off from the nuances of oppression in urban life. It appears that the transformation of urban life is meant to be achieved by pragmatic thought and action that identifies openings in everyday life for interventions, be they popular mobilisation or policy wise (cf. Huchzermeyer, 2018). The right to the city can be a starting position or optic, amongst many others, whereby such openings are identi-fied and acted upon. The objective of intervention is not simply greater and broader access to public spaces. Intervention also requires an ability of diverse members of society to shape public spaces, making them their own and more broadly func-tional for society conceived of in broader terms (Harvey, 2008:23).

The laager facilitates narrow conceptions of what is considered 'normal' in a democratic South Africa. These notions are often inflected with identity politics and attempts at restoring or preserving a reiterated state of apartheid by portraying fellow residents as matter out of place. Neighbourhoods become sites of assumed purity, loosely sutured around ethnicity, class or otherwise, that need preservation. Aspects potentially contravening 'acceptable' human existence are often mentioned. These include drinking during daytime in the week, cleanliness or lack thereof, and being black. Those (vaguely) deemed to be the devious others may apparently be harassed.

Crime as troubling the laager

The laager is frequently troubled, through crime and issues of municipal service delivery. Service delivery is in a sense viewed by members of the economic elite as being controlled by *the* political opponent constructed through logics of equivalence. The frustration of this part of the public is exacerbated by the fact that the newly formed JB Marks municipality has gone bankrupt since the merger of the City of Tlokwe Local Municipality and the Ventersdorp Local Municipality, as the new city council had to take on the latter's significant debt.

However, even in light of the fact that Potchefstroom has much crime, there are numerous other infrastructures that work to protect those living inside the laager. This may include the SAPS. Consider the following:

> Warning from SAPS: Please ensure that your vehicles are INDEED locked. The persons who use remote controlled blockers are again very active!! Also ensure that you close gates and are not hindered by rocks, etc.
>
> (Author's translation, JB Marks Security Facebook group,
> 22 September 2017)

Middleclass households also have access to private security services and infrastructures such as alarms, electric fences, high walls, security gates, and sometimes the consequences of calls for solidarity and shared interests in preserving a particular status quo. As the following quotations suggest, these might include neighbourhood watches and social media such as WhatsApp groups to report crime or even 'alert neighbours'.

> Please be cognisant that video material was received this afternoon of attempted theft of dogs from a yard in the Jeugd/Wandrag Street area. Observant neighbours stopped the process.
>
> (Author's translation, JB Marks Security Facebook group,
> 17 May 2017)

> Some or other criminal tried to enter my yard last night while my wife was home alone. Thank you very much to (name of security company) security and Afriforum neighbourhood watch for your excellent, fast reaction.
>
> (Author's translation, JB Marks Security Facebook group,
> 8 January 2017)

Afriforum is a large civil society organisation that has been associated with campaigns for Afrikaner interests, even though they often profess that this summation is inaccurate. This organisation has its own neighbourhood watches that run parallel to those sanctioned by the CPF.

Theft of motor vehicles may at first glance be viewed as troubling the laager, but this inconvenience is simply that: an inconvenience. Owners, who reside in this part of the JB Marks municipality, typically have insurance and can very soon resume their regular daily activities. Theft from motor vehicles, especially while the owner is in the car (i.e. robbery by definition), is a more sobering experience and one that has been reported on various occasions. Housebreaking and house robberies (when residents are home) are similarly sobering. It is not only that material wealth is threatened and that objects of sentimental value are taken away. To reiterate a previous point, it is also that specific forms of domestic life are threatened when the citadel is breached. Here, the limits of closing the frontier through abandonment, frontier governance and the control residents have over their own well-being become apparent.

Beyond the materialised laager, there are parts of the city where less affluent residents must sometimes travel. These areas are often the topic of complaints and reports of robberies and theft from vehicles. The area around the taxi rank and *Wandellaan* (pedestrian street) is a case in point. This area has been the subject of various official police warnings to the public for high incidents of theft from vehicles and robberies. The *Wandellaan* includes a host of budget furniture stores, while streets close by such as James Moroka, house various shops. The taxi rank is next to the River Walk Mall, which, subsequent to the development of the Mooiriver Mall, is colloquially known as the 'black mall'. It is a shopping centre more catered to less affluent households. It houses various types of budget retail stores and micro-credit providers, including 'specialist' branches of major financial institutions. High interest micro-loans may also be viewed as a tool of closure through abandonment. Poor citizens are subjected to what amounts to prolonged dependence in unending debt cycles, while a micro-loan provider and their shareholders benefit. Affluent residents can generally avoid these areas, should they so choose. Now though, we must conclude by considering means of constructively troubling the laager and reiterating what has been said about the relevance of the laager for those who study crime and security.

Conclusion: constructively troubling the laager

This chapter has argued that the laager might be more usefully conceived of and operationalised as a piece of infrastructure instead of merely a mentality. By doing so, researchers may unearth more precise and nuanced insights into the everyday circulation of ideas and practices, and their significance. The chapter has also presented the argument that for some the fear of crime is intertwined with discourses of onslaught rearticulated since the days of the total onslaught in the 1980s. This reiteration is somewhat different from that framing of the concept.

Some of the old discourses associated with the total onslaught appear (such as a pervasive mystical (criminal) insurgent) and justifications for (quasi)militarisation, in this case the militarisation of the suburbs. Such a view of crime, by which additional divisive discourses are attached to the reality of crime, facilitates retreat and insularity based on notions of purity and matter out of place. Victimhood is often linked with the securitisation of public spaces, which, in the context of a metaphorical (re)opened frontier, render them ripe for recolonisation by the few. The laager has been discussed as a key piece of infrastructure in this process: a set of physical and ontological structures. Recolonisation, taking place side by side with semigration, in a sense extends the laager, creeping and encroaching upon the spaces afforded to the general public for equal enjoyment. I have shown how the penetration of criminal activity troubles the laager, but this is not a productive way of troubling this infrastructural collective, which is often problematic because of some of its constituent parts. I will therefore conclude by considering briefly the potential for more constructive ways of troubling the laager in an attempt to gradually move away from a frontier mentality and its inherent practices of border-making through laagerisation. Floating signifiers in the Laclauian sense such as the victimhood of the privileged, purity and matter out of place, need to be challenged and exposed for their counter-democratic nature. A few examples cited in the following might provide preliminary points of entry for such progressive praxes. Other floating signifiers and strategies for harnessing their potential with them will be highlighted throughout the remainder of this book.

In the first instance, and notwithstanding the arguments presented here, social media users are not a completely homogenous group. There is sometimes also praise for the police. The following example is a case in point:

> A few weeks ago, I asked where the police are with the drugs and street women. Well, I just want to say well done to adjunct officer Stemmet and colleagues of the Potch Police!! Well done and thank you!
>
> (Author's translation, Potchefstroom CPF Facebook group,
> 25 August 2017)

Herein potentially lies a basis for constructive engagement. Initiatives from the police and the city council aimed at improving relations with this particular sector of the Potchefstroom public could set in motion a cumulative and iterative process of garnering support from those with more inclusive perspectives. This might spawn greater collaboration in addressing the crime problem and incrementally influence the perspectives of more conservative groups.

Similarly, my fieldwork with security companies revealed that armed responders are not a uniform group. Many are firmly aware of and respectful of the rights of others. For example, in one instance I observed the obvious embarrassment of an armed responder because he had to interrogate a person who only somewhat fitted the description of a suspect in a robbery. Armed responders need a plausible reason for why they cannot always abide by the at times unreasonable demands of clients. Although legislation does not often help to address real-life problems, perhaps this is a case where legislation (new or existing) that is enforced can provide armed responders with a valid reason to act more in keeping with democratic principles. Educating the general public about their rights when confronted by PSCs might also be of use.

Fundamental to the problematic described in this chapter and notwithstanding the rise of a political elite, is the relationship between race, class and geography, and the associated ideas of purity, danger and matter out of place. There are currently opportunities to disrupt this relationship. For example, in 2018 a vendor parked a caravan next to school in an affluent suburb from which people who do not reside in the area are selling food. There were complaints from residents about this, with vague accusations of it posing a threat of crime, and of course, a threat to property values. Luckily these complaints were not sufficient to remove the caravan-shop. At least a significant number of residents and scholars are acting as a willing market for the produce sold in their neighbourhood.

Another interesting example is the planned low-cost housing scheme on the *Vyfhoek* plots, a middleclass to upper-middleclass area. There have been various complaints on social media about the perils of such a settlement, very similar to those about the caravan-shop. For similar reasons, it is important that these types of projects are not prevented. As many have a strong commitment to insularity, it is perhaps necessary that integration across race, class and other identities at times be enforced. In other words, there needs to be more examples where more open and inclusive representations of space as per the work of Lefebvre (1991) hold sway over insular spaces of representation. This may reduce some of the ignorance evidenced in this chapter about the realities lived by fellow citizens. It could at least remove plausible deniability in this matter.

This chapter has also argued that the laager as an analytical tool might, depending on context, have advantages above notions of semigration and fortresses. The laager is particularly useful in the context of an infrastructural analysis, as, as I have defined the concept, it combines the physical and ontological. In this way the concept of the laager aids analyses that aim to better understand the flow of ideas, practices, objects and people associated with the politics of crime and security. Stated differently, although there is much evidence of the laager as a set of attitudes, the material and imagined reality of the laager enables unhelpful articulations. Unlike other concepts such as semigration and fortress, the use of the laager as conceived of in this chapter also helps move our analyses beyond gated communities, in a way that is still coherent.

The stigmatised spaces mentioned earlier, such as the *Wandellaan* and taxi-rank, are the topics of the following chapter. It is argued that these spaces along with other parts of the open frontier provide a necessary constitutive outside for the laager, by which increased policing inside the laager and the laager itself can be justified.

References

Ballard, R. 2004. Assimilation, emigration, semigration and integration: White peoples' strategies for finding a comfort zone in post-apartheid South Africa, in N. Distiller and M. Steyn (eds.) *Under construction: Race and identity in South Africa today.* Sandton: Heinemann, pp. 51–66.

Bigo, D. and McCluskey, E. 2018. What Is a PARIS approach to (In)securitization? Political anthropological research for international sociology, in A. Gheciu and W.C. Wohlforth (eds.) *The Oxford handbook of international security.* Oxford: Oxford University Press.

Blakely, E and Snyder, M. 1997. *Fortress America: Gated communities in the United States.* Washington, DC: Brookings Institution Press.

Botha, C.B. 2008. South Africa's total strategy in the era of Cold War, liberation struggles and the uneven transition to democracy. *Journal of Namibian Studies*, 4, pp. 75–111.

Buzan, B., Weaver, O. and De Wilde, J. 1998. *Security: A new framework for analysis.* Boulder: Lynne Rienner Publishers.

Cozens, P.M., Saville, G. and Hillier, D. 2005. Crime prevention through environmental design (CPTED): A review and modern bibliography. *Journal of Property Management*, 23(5), pp. 328–356.

Douglas, M. 1966. *Purity and danger.* London: Routledge.

Harris, B. 2002. Xenophobia: A new pathology for a new South Africa? in D. Hook and G. Eagle (eds.) *Psychopathology and social prejudice.* Cape Town: University of Cape Town Press, pp. 169–184.

Harvey, D. 2008. The right to the city. *New Left Review*, 53, pp. 23–40.

Huchzermeyer, M. 2018. The legal meaning of Lefebvre's right to the city: Addressing the gap between global campaign and scholarly debate. *GeoJournal*, 83(3), pp. 631–644.

Jürgens, U. and Gnad, M. 2002. Gated communities in South Africa: Experiences from Johannesburg. *Environment and Planning B: Planning and Design*, 29(3), pp. 337–353.

Laclau, E. and Mouffe, C. 2014[1985]. *Hegemony and socialist strategy.* London: Verso.

Lang, R.E. and Danielsen, K.A. 1997. Gated communities in America: Walling out the world? *Housing Policy Debate*, 8(4), pp. 867–899.

Learning English. 2017. Words and their meanings: Circle your wagons. Online: https://learningenglish.voanews.com/a/words-and-their-stories-circle-your-wagons/3860350.html. Date of access: 21 March 2020.

Lefebvre, H. 2000. *Writings on cities.* Translated by Eleonore Kofman and Elizabeth Lebas. Malden: Blackwell.

Lefebvre, H. 1991. *The production of space.* Translated by Donald Nicholson-Smith. London: Blackwell.

Lemanski, C. 2006. Residential responses to fear (of crime plus) in two Cape Town suburbs: Implications for the post-apartheid city. *Journal of International Development*, 18(6), pp. 787–802.

Low, S. 2001. The edge and the center: Gated communities and the discourse of urban fear. *American Anthropologist*, 103(1), pp. 45–58.

Nauright, J, 1996. A Besieged tribe? Nostalgia, white cultural identity and the role of rugby in a changing South Africa. *International Review for the Sociology of Sport*, 31(1), pp. 69–86.

Oxford Living Dictionary. CV. Laager. Online: https://en.oxforddictionaries.com/definition/laager. Date of access: 9 July 2018.

Paasche, T., Yarwood, R. and Sidaway, J. 2014. Territorial tactics: The socio-spatial signifi-cance of private policing strategies in Cape Town. *Urban Studies*, 51(8), pp. 1559–1575.

Posel, D. 1984. Language, legitimation and control: The South African State After 78. *Collected Seminar Papers, Institute of Commonwealth Studies*, 33, pp. 139–151.

Shearing, C and Wood, J. 2003. Nodal governance, democracy and the new 'denizens'. *Journal of Law and Society*, 30(3), pp. 400–419.

Soudien, C. and Sayed, Y. 2004. A new racial state? Exclusion and Inclusion in education policy and practice in South Africa. *Perspective in Education*, 22(4), pp. 101–115.

South Africa. 1996. *The constitution of the Republic of South Africa. Act 108 of 1996.* Pretoria: Government Printers.

Statistics South Africa (StatsSA). (2017). Tlokwe City Council. Online: www.statssa.gov.za/?page_id=993&id=tlokwe-city-council-municipality. Date of access: 31 July 2017

UrbanDictionary.com. Cv. To circle the wagons. Online: www.urbandictionary.com/define.php?term=Circle%20the%20Wagons. Date of access: 21 March 2020.

UsingEnglish.com. To circle the wagons. Online: www.usingenglish.com/reference/idioms/circle+the+wagons.html. Date of access: 21 March 2020.

Interview cited

Private security company manager. 2018. Interviewed by Gideon van Riet. 6 June. 2018. Potchefstroom.

Facebook posts

The full details for each post is supplied in each in-text reference. These references include enough detail for interested parties to confirm the citation. However, supplying additional information here, would mean linking comments/quotations to names, which would contravene the conditions set by the group administrators for usage of this data.

3 Infamous spaces and the constitutive outside

Introduction

This chapter completes Part I of the book, thus completing the conceptual foundation for what remains. The chapter is about the opposition between notions of inside and outside. We may trace the argument in this book thus far to better explain how this chapter reveals active border-making through such inside/outside equivalences. This book draws on an analytical approach that focusses on the flows of ideas, objects and people directed by and through security infrastructures. The context of Potchefstroom is located within the context of contemporary South Africa, where the aggregate of social orders is characterised by a reopened frontier. In this context, hegemony is directed, firstly around the notion of closure through abandonment. Secondly, frontier governance and border-making through laagerisation, or instilling a collective of physical and ontological structures that govern spatial difference, are the main modes of governing. The research also draws on a Laclauian notion of post-structural hegemony and to a lesser extent on Lefebvre (1991), especially his dialectic between official representations of space and spaces of representation, to make sense of, and critique the flows directed through security infrastructures, and as such some of the mechanisms of hegemony. The study has been conceived of as moving beyond discourse analysis to also understand the relationship between discourse and space. In this regard, the notion of a constitutive outside located within the Laclauian dialectic between time and space is instructive (cf. Laclau, 1990:19). Spatial configurations, as a function of hegemony remains dynamic, although change might be slow and non-linear. Thus, to reiterate, the Lefebvrian ideas used are subsumed under a Laclauian epistemology.

Given this outline, this chapter engages with border-making practices in Potchefstroom through the notion of stigmatisation. Stigmatisation is a key tool through which the relationship between the non-continuous laager and its constitutive outside is mediated. We may view stigmatisation as a tacit and/or deliberate tool of biopolitical abandonment. The chapter engages with the oppositions that constitute the outside of the laager through stigmatisation. In the process, it considers the adverse consequences of such stigmatisation.

Methodologically, the chapter provides a theoretical overview of Wacquant's work on territorial stigmatisation and its relevance to South Africa. The chapter

DOI: 10.4324/9781003028185-5

then draws on data from the local *Potchefstroom Herald*, social media platforms, interviews with diverse actors, including estate agents and citizens and notes from participant observation to explain and critique stigmatisation and border-making in and around Potchefstroom. These pieces of data are brought into conversation with the burgeoning literature on xenophobia in South Africa. The purpose is to explain what spaces are stigmatised and in the process form part of the constitutive outside of the laager, and to understand how stigmatisation takes place and what its effects are. In the process the analysis highlights float-ing signifiers that might serve as points of opposition to aspects of the dynamic hegemonic status quo.

A constitutive outside and stigmatisation

Laclau (1990:19) notes that no subject or force is self-constituting. Identities are always relational and as such constituted through that which is outside of it. The non-continuous laager as a set of physical and ontological security infra-structures that characterise an approximate territory has an outside or a set of outsides. This outside, for our current purposes, is largely composed of partially or more fully abandoned spaces and populations. These spaces include stigma-tised spaces, especially closer to the dynamic perimeter rendered through security infrastructures. There is a linguistic and a spatial component to this dynamic. If as Laclau (1990:41) argues time and space are in a dialectic relationship, then the contestation of borders between the laager and its outside is arguably where the *politics* of crime and security is most significant. There is a circular relationship between these two elements. Reducing them to independent and dependent vari-ables would arguably miss the point, which is to find instances (articulations) for disrupting the laager. To be more precise, the laager is significantly articulated through stigmatised spaces and stigmatised spaces are largely constituted through a conspicuous silence on the reality of the laager. This makes stigmatisation, if not vital to, at least extremely important in the process of laagerisation. Contemporary hegemonic formations thrive off of stigmatisation to justify the laager, while not acknowledging the latter as a collective instrument for border-making. Therefore, the most significant silence in a sense pertains to the inside and not the outside of the laager. I do not imply that this inside/outside divide is clear-cut or that the exact nature of the border is not open to interpretation. Of course it is, and this chapter will demonstrate how border-making is dynamic. I do, however, argue that this division between inside and outside and the logic whereby it is sustained is real.

It is through stigmatisation that symbolism meets political economy in border-making. Slater (2015:113) notes that of the two broad schools in Urban Studies, human ecology and political economy, neither satisfactorily deal with symbolism. The work of Waquant (cf. 2007, 2008, 2016) is useful in introducing this symbolic component to our understanding of the urban, although, as with any conceptual scheme developed in the Global North, this perspective cannot be adopted uncriti-cally in southern contexts. Some caveats do apply to adopting Wacquant's work in the Global South and in particular the spatial-temporal realities of contemporary Potchefstroom.

Central to Wacquant's work is the concept of advanced marginality. This phrasing in itself suggests a logic designed around experience in wealthier countries. There are however dimensions to the concept, that may be elucidated and adapted for the purposes of this chapter. These dimensions relate to the political and economic foundations of marginality and its extension through the symbolism of stigmatisation, which is referred to as symbolic defamation (Slater, 2015:113). Finally, the concept of advanced marginality is associated with certain adverse consequences for inhabitants of stigmatised spaces.

There is much similarity between the political-economic foundations for advanced marginality highlighted by Wacquant and the analysis of the previous two chapters. Wacquant talks about the post-industrial city in an era of neoliberal ascendancy. These relegate some to particular socio-spatial formations (Wacquant, 2016:1077). Here the links to global transformations over the past 50 years and the link to the ubiquitous, though often dangerous concept of neoliberalism, are clear. Wacquant, however, offers enough detail to let the reader know what he means by 'neoliberalism', that is, what his concerns are with large-scale transformations in recent decades. The duel moves in advanced economies from Keynesianism and Fordism to individual responsibility for welfare and Post-Fordism, rendered a political economy where similar, though not necessarily identical, forms of abandonment of segments of society have occurred. The result has been that they were forced into the cheapest accommodations. Their neighbourhoods suffer the consequences of disinvestment and disproportionately high levels of unemployment. These parts of the city are prominent sites of social housing, but also much neglect by authorities. Gentrification often replaces disinvestment. This often forces people out of properties they can no longer afford (Slater, 2015:119). In these areas greater punitive measures accompany greater social insecurity. Financialised capitalism contributes to and interacts with an unskilled labour surplus and thus precariousness (cf. Wacquant, 2016:1082).

According to the logic of advanced marginality, the term 'prison fare' describes an internal feature of neoliberal regimes. The police, the courts and prisons are important political institutions that actively produce and manage and maintain inequality and marginality (Cummins, 2016:77). In these contexts, we find

> forms of poverty that are neither residual, nor cyclical or transitional, but inscribed in the future of contemporary societies insofar as they are fed by the ongoing fragmentation of the wage labour relationship, the functional disconnection of dispossessed neighbourhoods from the national and global economies, and the reconfiguration of the welfare state in the polarizing city.
>
> (Waquant, 2007:66–67)

As for the political-economic foundations of advanced marginality, there are some commonalities and differences with the general South African context as explained in previous chapters. Large-scale market deregulation has a shorter history in South Africa. The consequence has, however, been that the post-apartheid society has become *materially* more unequal than the apartheid society. South Africa has also had other forms of dispossession in its history, that were more blatant and provided a platform for certain types of deregulation to have a devastating compounding effect. Unlike North-Atlantic contexts where the distinction between social and other housing or between social, middle-class and working-class housing is used, South Africa has at least four discrete classes of housing. In South Africa there is an additional category and significant component of informal housing, where social housing in the form of the RDP houses has not yet materialised. As explained in the following, these spaces often suffer special stigmatisation. While not all of Wacquant's notion of advanced marginality is precisely applicable to the South African context, the notion of a 'topography of disrepute' is a valuable concept, precisely because it can be used to differentiate between diverse contexts within a more general social order.

The symbolic component of advanced marginality relates to the popular association of these areas with pathologies and social problems. There is a constitutive power to symbolic structures manifested in 'institutions and systems of dispossession as they are internalised by actors' (Wacquant, 2016:1084). Populations residing in these areas are associated with the abject, which justifies punitive measures (Tyler, cited by Slater, 2015:118). Wacquant et al. (2014:1270) explains,

> Territorial stigmatization is not a static condition or a neutral process, but a consequential and injurious form of action through collective representation fastened on place.

Wacquant offers a counter to dominant narratives which renders pathological the poor and the places where they reside (Wacquant et al., 2014:1271). Therefore, neighbourhoods are not a manifestation of problems to be addressed, they are articulated as being the problem itself (Slater, 2015:123). Society individualises poverty, as I have argued in previous chapters they do this with crime, instead of acknowledging the social and structural causes of poverty and crime (Cummins, 2016:78). Ideally one would have a more nuanced reading of the social dynamics, which produce poverty and 'deviance', that acknowledges structure and agency. Stigmatisation, however, only emphasises individual agency or the lack thereof. In the context of South Africa, where nearly 40% of working aged people are not formally employed, such stigmatisation of a lack of productive agency is as inexplicable, as it is prevalent. All these people cannot possibly be too 'lazy' or 'stupid' to be productive. Conversely, we might also question the implied assumption that employed people are necessarily hard working. In the South Africa context, stigma latches onto established lines of division, such as class, race, ethnicity and citizenship. Foreign nationals too, are often stigmatised, and as such, become integral to the stigmatisation of spaces. Although some foreign nationals, just like many South Africans are involved in crime, this is not universal. Moreover, as will be argued, crime remains, at least significantly, a function of limited available livelihood options for immigrants, just as it is for South Africans.

Wacquant et al. (2014:1275) distinguish between five effects of territorial stigmatisation. Residents of these areas lose a sense of self. Their social relations are harmed and they cannot act collectively. They learn to cope through means that tend to validate the stigma. Occupants of surrounding areas actively avoid these areas and their residents and employers discriminate based on 'address'. Municipal service delivery to these areas is neglected, except for policing and surveillance, which is much higher than in other parts of the city, where it would not be accepted. Symbolic production in the work of journalists, scholars, policy analysts and politicians on the area breeds more similarly denigrating symbolic production and hegemonic articulations. Finally, stigmatisation informs the beliefs, views and decisions of state officials. This in turn influences public policies that combined with 'market and other forces, determine and distribute marginality and its burdens'.

Not all of this seems relevant to the South African context and in particular Potchefstroom. Most significant for our immediate purposes are the observations about policing. In Potchefstroom, it is mostly the suburbs that are consistently patrolled, by PSCs, with significant approval and funding by middleclass residents. This type of policing and surveillance is a key part of border-making and maintenance. Far less clients of PSCs live in these stigmatised areas, as the costs of a subscription are prohibitively high. Because residents of, and businesses operating in, stigmatised spaces depend on the SAPS, the extent of security provision they receive is limited by the capacity and efficacy of that organisation. These communities are to a large extent left to their own devises and as such are – for the most part – abandoned outright, and not in combination with more draconian policing. The policing mostly takes place in the middleclass suburbs, although it is mostly policing against those who reside in stigmatised spaces. As such, border-making is important in sustaining the contemporary dynamics of hegemony. Some infrastructures, such as PSCs, the *Potchefstroom Herald* and social media platforms circulate discourses and practices that sustain hegemony through laagerisation. Consequently, the dynamic laager is sustained through modes of frontier governance that emphases notions of difference based on logics of equivalence. We must now turn to some concrete examples to illustrate these dynamics of stigmatisation, border-making and laagerisation.

The central business district

If I were to leave my flat in the fringes of the Potchefstroom CBD, turn to the right and walk about 500 m in a southerly direction, I would reach the currently infamous *Wandellaan* or pedestrian street. The *Wandellaan* is found where the street I live in turns into a one-way street. As such, the name is a misnomer, as it is not a pedestrian walkway exclusively. The street is surrounded by what many middleclass people view as the seedy underbelly of the city. This is a popular view, though it is ignorant on many accounts. There is much property crime in this area (Anon, 2018:4, 3 May). Theft out of motor vehicles takes place here quite often. In recent years, there have been more muggings in the CBD than in many other parts of the city. However, if we are talking about levels of fear then this example of the CBD and the second example, Bult-West are not nearly as significant as the outlying extensions of Ikageng. Violence is a far more frequent occurrence in those parts. Still many prejudicial imaginaries have been projected onto the *Wandellaan* and the surrounding CBD. These imaginaries pertain to familiar racial discourses associated with South Africa, but also *more recent* (though not recent, as such) anti-immigrant sentiments. According to popular beliefs, this is where 'The Nigerians' live, even though many immigrants of Nigerian descent, including 'The Nigerians' referred to, live in various other parts of the city. This type of stigmatisation relates to what Shaw (2002:291) refers to as 'the Nigerian problem' in his critical reflection on the development of West-African criminal networks in South Africa. 'The Nigerians' refers to drug dealers. But, 'Nigerian' has also become a collective name for immigrants from all over Africa (Van Riet, 2020:91). Such a logic of equivalence, which renders diverse groups into the single category of drug dealers is highly problematic.

It is also in this part of the city where 'The Pakistanis' work, another collective term for recent immigrants from South Asia, also including Bangladesh and even parts of North-East Africa. All of these residents supposedly own dodgy cell phone repair shops that serve as fronts for buying and selling stolen goods. Again, while this is possibly true for some recent immigrants from Asia, it is unlikely to be true for all of them. The CBD is also where, possibly desperate, South Africans steal and sell products to foreigners. Middleclass people, based on social media analysis, tend to encourage each other to avoid this part of the city. The CBD has been neglected as new developments aimed at wealthier residents have been developed elsewhere.

In 2008 the Mooi River Mall was opened opposite the old River Walk Centre, on the edge of the CBD and next to the taxi rank where minibus taxis pick up and drop off residents of Ikageng and surrounding areas who come to Potchefstroom to work or shop. The River Walk Centre includes a major budget grocer, named Shoprite while the new mall features higher end retailers and Checkers, the upmarket version of Shoprite Holdings' suite of stores. As explained in the previous chapter, soon after the completion of the Mooi River Mall, the nature of the River Walk Centre changed. It became a site for micro-credit providers and retailers of lower end goods.

Furthermore, with regard to security infrastructures, the institution of car guarding for the Mooi River Mall works different compared to most other parts of the city. Car guards are a South African phenomenon. They typically work near any commercial setting and at shopping centres. Once you park your car, they would ask you if they may look after it while you are away. If you say yes, then you give them a small donation when you return. The amount of the donation is up to the car owner's discretion. Traditionally car guards have been

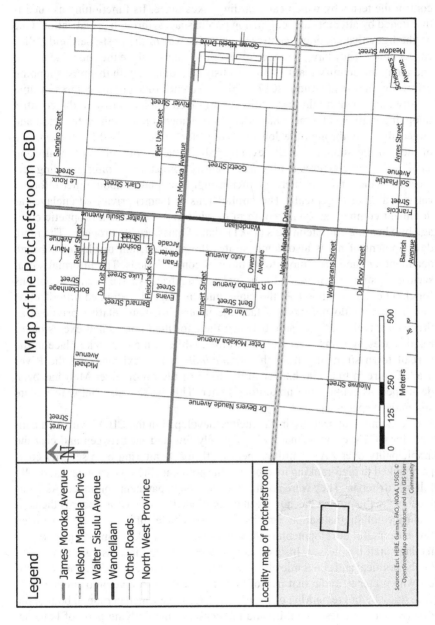

Map 3.1 The central business district

self-appointed. The practice may be viewed as an informal social security system, in light of significant unemployment. The Mall is on private land, similar to the West Acres Centre in another part of the CBD. Management can therefore control the terms by which car guarding takes place. Its functioning as such is threatened by attempts at recolonising public spaces, through privatisation. One might argue that shopping centres are a major part of the post-apartheid political economy. They have sprawled all over the country since the end of apartheid and today the country features high on lists of countries with the most shopping centres in the world (Anon, 2017). This prevalence of shopping centres is quite bizarre and, some might argue, distasteful given the inequality in the country. Shopping mall development has offered investment opportunities for retail and larger investors through the Johannesburg Stock Exchange (JSE). Indeed, one of these listed shopping mall developers built the Mooiriver Mall. Malls and smaller shopping centres tend to hire security companies to offer guarding services in a context where there are increasingly complaints about other informal car guarding arrangements. The implications of (semi) privatised enclaves as dominant commercial zones conform to what now seems as a prophetic coinage by Shearing and Stenning's (1981), that of mass private property. The rapid development of malls also demonstrates the phenomenon of closure through abandonment and once more to its internal contradictions. The informal social security system is eroded and marginal South Africans have even less options for eking out a livelihood. At the same time the laager is extended. Middleclass South Africans do not have to face the inverse of their relative privilege as they can travel relatively safe between their fortified homes and exclusionary retail spaces by using their cars. Therefore, while crime does take place in the part of town stretching from the *Wandellaan* to the taxi-rank and the River Walk Centre, a financial barrier in the form of the Mooi River Mall has been developed to contain the ubiquitous Other. This barrier also helps to isolate stigmatised spaces.

One example of security infrastructure developed in the CBD involved a contract for CCTV cameras that were relatively few and far between and cost the municipality over ZAR 1 million per month for monitoring by a security company linked to high-ranking members of the political elite. The cameras gradually fell into disrepair. They were not fixed, even though payment continued. As far as I am aware the project has now been discontinued. However, as regards the consequences of stigmatisation, smaller businesses have suffered. There have been few meaningful developmental interventions aimed at addressing crime and protecting small businesses. Instead, the CBD has become increasingly abandoned by the political and economic elites, as it struggles to attract middleclass customers who prefer to avoid that part of the city. I will now move onto another brief case study before remarking of the popular discourse which pertain to both of the first two case studies. The third and final case, of the outlying parts of Ikageng, pertains to somewhat other social economic and political dynamics and will be dealt with subsequently.

Bult-West

The term Bult-West was first identified by an estate agent in an interview (Interview, April 2018), when asked which areas that are potentially at risk of losing tenants. This part of the city covers mostly student accommodation towards the North West of the city, but also cheaper accommodation on the West of the centre of the city. It is essentially found between the railway line, Chief Albert Luthuli Street and Louis Le Grange Street to the West. The latter includes a large immigrant population and indeed, it would appear a major node for drug dealing, although drug dealing takes place basically anywhere in the city. The *Herald* has repeatedly reported on a particular house in this part of the city, with CCTV cameras pointing to the street, where drugs have been sold and from where deliveries are apparently dispatched. This area is surrounded by more affordable student accommodation and blue-collar homes.

Then moving to the North, increasingly in recent years the *Potchefstroom Herald* and social media has reported armed robberies of students, mostly returning home in the early hours of the morning after a night out or after working until very late in a university computer laboratory (cf. Wetdewich, 2018:3; Wetdewich and Saayman, 2018:3). Certain street names are repeated over and over in reports on crime. But, what makes this part of the city different from the historically working class community to its South is the extent of coverage in the media. This additional coverage might be because robbery is considered a more traumatic crime than drug sales. I would accept that argument, although interview data in both Potchefstroom and Ikageng generally suggests the two crimes are often linked. This additional coverage of robberies might also relate to how important these two parts of 'Bult-West' are to the community in relative terms. Finally, reporting in the *Herald* and other media, almost certainly has a lot to do with reporting to the SAPS. Robberies are much more frequently reported than drug sales, for the obvious reason that the parties involved in drug sales have no interest in the police knowing about their activities.

The response to crime is however most telling. There have been clear attempts to actively combat stigmatisation in the northern part of Bult-West, to the North of Chief Albert Luthuli Street. This urgent response may relate to the university's purported image as a safe environment and, I would argue, with preserving the identity of sections of the city. There has been an active move to reduce stigma. The university has long maintained that it is a safe alternative to other universities in the country. Moreover, while the South has always been more of a blue-collar area, the northern part of the city has always been wealthier. Initially, much of the Bult area was a wealthy lower density residential neighbourhood. Over the past 20 years, numerous houses have been demolished to build blocks of flats for student accommodation, as the university's student numbers increased. The result on increased crime on the Bult broadly speaking, has since 2019 been the development of the Cachet Park CID, which is the focus of Chapter 9.

Thus we have two parts of the city close to each other, literally separated by a road, both of which have suffered stigmatisation. One of the two sites, however, has seen an active fightback to remedy the situation, while the other has more or

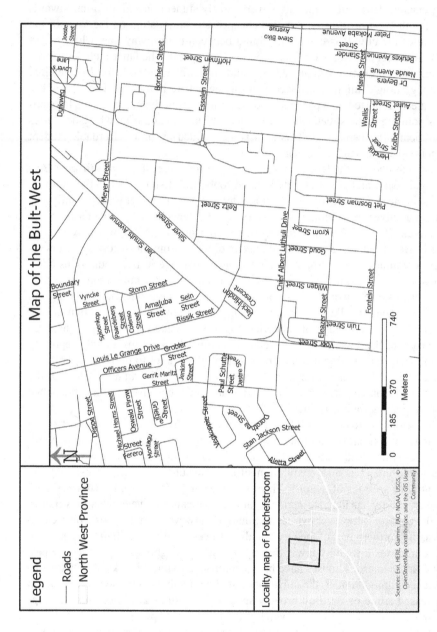

Map 3.2 'Bult-West'

less been left to its own devises. The distinction brings up questions regarding why some spaces, under threat of stigmatisation are simply left and not overly policed in the ways Wacquant explains, while the other has been subjected to increased private policing, not so much to punish as Wacquant has found elsewhere, but rather to reduce crime, protect an image and possibly also property values and the sustainability of the university. In both case studies discussed thus far, pertaining to the CBD and Bult-West it has been private and semi-state actors that have been the impetus behind enhanced security in the wake of stigmatisation. Hence, it would appear that some types of practices and spaces are more important to the contemporary hegemonic order than others. Now we should consider in greater depth the stigmatisation of foreign nationals as evidenced in the case studies mentioned earlier, by drawing on the existing South African literature on xenophobia.

Xenophobia

South Africa has long been a destination for migrants from southern Africa, working on the mines, as part of the migrant labour system (cf. Wolpe, 1972). More recently it is not the migrant workers from Malawi, Lesotho and so forth who work on the mines, who are significant in public discourse. Hostility towards immigrants from other African countries have been reported from the 1990s (cf. Morris, 1998) and has continued to this day. With democratisation South Africa became a relatively popular destination for immigrants from other African countries. These immigrants included professionals, students and refugees from political persecution or because they too struggled to find employment in their respective countries of origin (Isike and Isike, 2012:93; Morris, 1998:1120). Yet, there has been a persistent association between the previously mentioned category of 'The Nigerians' and drug dealing. Various authors have been critical of this particular logic of equivalence (Adesina, 2019; Isike and Isike, 2012; Neocosmos, 2008:588; Morris, 1998:1120). The resentment of immigrants amongst other reasons, for a supposed association with crime is not unique to Potchefstroom (cf. Adesina, 2019:109).

Many explanations have been offered for these attitudes towards African migrants in South Africa. Matsinhe (2011) draws on psycho-analysis and the inferiority syndrome that leads to hate of the self and those who physically most resemble that self. This loathing eventually manifests as a type of projected hatred onto the new 'other'. There is a social unconscious or the stock of prejudices, myths and common-sense knowledge (Matsinhe, 2011:300). To these ideas which mostly explain black on black hatred or even violence, Neocosmos (2008) adds a few other explanations. He argues that

> there is a hegemonic notion of exceptionalism in South African public culture, maintained by all, not only whites. The prevalent idea here is that 'the country is not really located in Africa.
>
> . . . and that its intellectual and cultural frame of reference is in the USA and Europe.
>
> (Neocosmos, 2008:590)

This exceptionalism is punctuated by the idea of an essential 'South African-ness' (Neocosmos, 2008:591). Potchefstroom is largely white and because of that the so-called 'bio-cultural hypothesis' presented by Harris (2002), seems quite appealing as an additional explanation for middleclass views in Potchefstroom. For many residents of Potchefstroom race as a stringent line of division remains true in a way analogous to the apartheid years. Like the laager in general, race still represents a type of 'apartheid imaginary'. This strict division is observable in various guises, including where the university is referred to by a white student as 'our place' (Wetdewich, 2018:5, 30 August).

None of the aforementioned arguments of course change the reality that the drug trade in Potchefstroom involves a significant number of residents of Nige-rian origin. However, I also have colleagues, who are full professors, who are originally from Nigeria. Indeed, as Isike and Isike (2012:110) note the number of Nigerians involved in criminal activities pale in comparison to the doctors, pro-fessors, engineers, architects, mathematics and science teachers and entrepreneurs from that country who live in South Africa. As Neocosmos (2008:590), however, notes the 'the politics associated with this discourse are invariably founded on the notion that migrants from Africa are here to take and not to give'.

South Africa is the largest drug market in sub-Saharan African Africa (Sicetsha, 2019) that includes a broad range of substances. This is clearly a problem in need of intervention. Residents of all races are involved as couriers and runners and according to some reports even higher up in the supply chain. Methamphetamine, Methcathinone, Methaqualone and MDMA (Ecstasy) production takes place in South Africa and based on interview data, at least some of these are produced in Potchefstroom. Meanwhile throughout South Africa, corruption of police offi-cials has been reported (Dlamini, 2019; Cheteni et al., 2018:13; Shaw, 2002:301), which has protected the industry. Many young users are caught up in the criminal justice system. This has knock-on effects, while targeting the illicit supply com-bined with a medical as opposed to a criminal approach to addiction might be more productive (Pinnock, 2019).

Importantly, South Africans as customers are also complicit. The drug trade in Potchefstroom remains robust. There appears to be a sense of 'not in my back-yard' (NIMBY), while students, and other rich and poor citizens support this industry. As such, there is a complex entanglement between the stigmatised and the 'normal' and between 'inside' and 'outside' of the laager. Those involved in criminality of this kind find it harder to settle in wealthier suburbs. They will be shunned. Still, interview data and newspaper articles (cf. Keppler, 2019), reveal that students and people from relatively well-to-do areas remain customers. The laager, in this context is simply a way of governing the NIMBY phenomenon by displacing the most obvious evidence of illicit trade. Therefore, the signifier 'Nigerian' should be reframed to distinguish between criminal networks, simply referred to as such, and migrants from Nigeria or elsewhere. To group permanent residents or naturalised South Africans under this discriminatory label is problem-atic as it breeds discord and conflict. The struggle is against crime and not against a particular defined people. As it stands the term reads and performs as a slur.

The same spread of educational level and economic activity in South Africa that counts for residents of Nigerian decent also counts for many immigrants from Asia often included under the banner of 'The Pakistanis'. Pillay (2010:40) reports that many immigrants in Durban are medical doctors, restaurateurs and other business owners. These residents typically have at least a high school education, but often also a university education. Many came to South Africa after better economic prospects (Pillay, 2010:41). In the Johannesburg CBD businesses are dominated by immigrants (Thompson and Grant, 2015). These businesses have frequently been targeted by authorities. 'Operation Clean Sweep' in Johannesburg led to the police shutting down many local and immigrant informal traders and businesses in the Johannesburg CBD. Often South and East Asian workers suffer the brunt of these attacks (Park and Rugunanan, 2009). Asian migrants remain vulnerable. They are targeted as foreign small business owners and extorted by the police (Park and Rugunanan, 2009:3). Market saturation has driven newer migrants out of major cities into smaller, cities, towns and townships (Park and Rugunanan, 2009:9). Potchefstroom and Ikageng have received numerous Asian immigrants in recent years and although some are reportedly involved in illegal activities, according to PSC interviews, once more the popular generalisation is inaccurate and potentially dangerous. They also suffer crime and according to Park and Rugunanan (2009:12) [a] 'number of Pakistani interviewees also accused Indian South African employers of exploiting desperate and destitute Pakistani and Bangladeshi migrants'. These observations corroborate some of the allegations I have heard during my fieldwork. It would appear that South Africans sometimes own spaces that they rent out to immigrants in exchange for excessive kick-backs or in other cases, where 'foreign owned businesses', often subjected to xenophobic attacks are in fact locally owned. In such cases the immigrants are simply employed because they provide cheaper labour and are exploitable. Again stigmatised foreigners are entangled with locals who escape stigmatisation. We may therefore problematise the signifier of the criminal outsider as being linked to an enabling insider as well as being a dangerous overgeneralisation.

Ikageng

Ikageng and the surrounding areas of Mohadin and Promosa are the most obvious stigmatised spaces outside of Potchefstroom. These spaces are, however, not stigmatised equally. In reality Mohadin as an 'Indian' settlement is very infrequently, if ever, publicly stigmatised. The informal segregation between white and Indian South Africans and between white, black and coloured South Africans remain more pronounced. It is especially Ikageng and Promosa that have suffered stigmatisation. Initially these settlements were the result of forced removals from the Potchefstroom (Jansen van Rensburg, 2006). Ikageng in particular has grown significantly since then.

Whereas previous examples of border-making have related to managing the inside of a dynamic laager, the case of Ikageng and surrounding areas is different.

These areas have always been on the outside of any articulated inside. Public stigmatisation merely reinforces that fact by portraying them as spaces of murder, rape, mob justice, criminal gangs and the root of housebreaking and robbery in Potchefstroom. Numerous articles on such crimes in Ikageng have been published in the *Potchefstroom Herald* in 2018 and 2019. They are far too many to mention here. Criminals in Potchefstroom are typically identified by race or by name, which in any event can be linked to race. This identification occurs both in the *Herald* and on social media. In all likelihood, an image of residents of Ikageng and black people more generally, as being violent criminals is perpetuated, even though Chapter 1 has revealed the social, political and economic dynamics of crime is more complex.

Pieces of formal infrastructure, following apartheid spatial planning, does a lot to maintain the laager. These include a railway line, a highway and most significantly an industrial park, purposefully developed, as was the case throughout apartheid South Africa, to maintain a spatial separation between black and white. Yet, black and white have always been entangled, even if only during working hours. Black South Africans throughout the apartheid and post-apartheid dispensations have served as cheaper labour on largely white owned mines, white owned farms and as housecleaners and gardeners in the suburbs.

Sites of stigmatisation are also linked. If we now return to the case of the CBD, the taxi rank has featured strongly in narratives of robbery, where suspects simply 'disappear into the taxi rank'. As far as crime is concerned the taxi rank is a potential source of criminals who move from Ikageng and back again after stealing goods in Potchefstroom. This densely busy square in the CBD offers great camouflage. Robbers would, the typical story goes, sell stolen goods in town and then leave, presumably by taxi. Robbery and theft is certainly one contemporary form of entanglement between different segments of the JB Marks municipality. On another front, the taxi rank is also a lifeline for the majority, as this and other less formal taxi pickup and drop-off points in the city offer means of access to people working in Potchefstroom who live in Ikageng. Many still work as domestic workers in middleclass homes. Given the levels of unemployment in the country housekeeping in itself might not be a problem. However, it does become problematic when wages are insufficient as is often the case in South Africa (De Villiers, 2020; Bangani, 2019). It should therefore be clear in this example too, that stigmatisation cannot rid itself of entanglement and as such the stigmatisation of black South Africans residing in townships like Ikageng is silent on how wage labour from places like Ikageng helps to construct the laager through 'efficient economics'. Lower wages may for example free up capital for security upgrades to a home.

Stigmatisation and other forms of border-making in Potchefstroom is reflected in the dialectic between representations of space and spaces of representation playing out, because Ikageng has reached physical saturation. The plots available for housing have been nearly exhausted. Ikageng is flanked by open land that belongs to the army and also land that is unsuited for development, because it is underlain by dolomite. Dolomite is a porous rock formation that disintegrates gradually when brought into contact with water. This in turn causes sinkholes and as such presents a risk to public safety. The plans are now to develop low-cost housing schemes, two of them as far as I am aware, on the Eastern fringes of Potchefstroom. The prospect of these developments has resulted in a public debate between the municipality and conservative constituencies. This debate is arguably the most obvious instance where representations of space through city planning have come up against more exclusive spaces of representation. In most cases the former has largely been absorbed or transformed by the latter, but in this instance, it would appear the municipality has dug in, citing the reasons stated earlier as the need to desegregate society (Leshage, 2018a:4, 3 May). While desegregation has long been an official objective of government and many would arguably have expected more to have been done by now, it is clear that for some the aforementioned developments represent a major disturbance in the order of things; in their ontological security. Thought processes and true intent is difficult to infer from the words and actions of people. It does however seem likely that the representation in the media of Ikageng and by extension black people, including an association with crime, informed some of the responses to these proposed developments. For some a lifetime of reinforcement through tangible and materially manifested, though not immediately obvious to all insiders, border-making likely also informed this response. Chapter 5 further explores some of these psycho-analytical aspects.

Discussion and conclusion

Although there is some merit in the concerns with crime in stigmatised spaces, not all residents and businesses who find themselves in those areas are problematic. Spatial stigmatisation, has implications for the very viability of small businesses that cannot afford the rent on the *Bult* and notably in the Mooi River Mall. It also has implications for property values and therefore the financial security of property owners. These actors, however, do not have the collective power to affect change that wealthier segments of society have.

This chapter has drawn on Wacquant's notion of territorial stigmatisation and essentially made an argument that fits in a gap between Wacquant's work and Foucauldian biopolitics. Stigmatisation and abandonment does not actively make die and let live in the pre-modern sense. It also does not 'make live and let die' in the modern sense. According to a Foucauldian reasoning sovereign power could 'take life or let live'. On the other hand, biopower fosters life or 'disallows it to the point of death' (Foucault, 2003:138). These are techniques of power that have changed over the centuries. Like disciplinary, biopower power circulates. It is indirect and not completely vertical, although the implication is not that the state and dominant classes are not key nodes the circulation of power. Rose and Miller (1992) in their paper on the governmentalisation of the state explain how the government and the state are intertwined with modes of governmentality and as such circulating power and for our purposes the flows through which hegemony is manifested. Biopower, disciplinary power, is also about populations and not only individual bodies. But, biopower does function through norms as opposed to overt rules. It is internalised by people and not exercised through the threat of violence, as lives are managed individually and/or collectively (Taylor, 2011:43–44). By revisiting the inside/outside distinction and the frontier as a collective mode of governing society, biopower becomes a bit more than merely 'letting' die. For most of the stigmatised spaces in Potchefstroom, and Ikageng, the hegemonic dynamic that permeates the social order and is not entirely directed from above, (in)advertently makes live *by* letting die. This is because the inside depends on its outside to secure itself. My argument is not so disagreement with Foucault, as it is a slight reformulation.

The political economic logic of closing the frontier *through* abandonment, self-contradictory as it is, implies that letting die is typically neither bloody violence nor is it simply leaving people to their own devices. Instead, closure through abandonment implies something in the middle, which we may call making live *by* letting die. The inside and outside as mutually constituting elements depend on each other to exist in their current approximate form. 'And' letting die underplays this dependence, at least if we consider contemporary readings of Foucault in the era of neoliberalism, such as Povinelli (2011) discussed in Chapter 1. In *Society must be defended* (2003:239, 241) he does mention racism and genocide and that the new right of making live and letting die does not replace the old sovereign power. Instead it infiltrates and permeates that power to reshape it. This racism, unlike the work of Arendt (1953) used in Chapter 1, is not (overtly) part of ideology. It is part of the prevailing technology of power whereby ideology (typically liberalism) and hegemony circulate (Foucault, 2003:258). As such, race is no longer a policy issue, at least not apart from (official) redress. Race does however pervade the status. It still implies the death of 'inferiors' for the benefit of 'society', but through the additional means of neglect and every other form of 'indirect murder' (Foucault, 2003:255–256). Somewhat differently for our purposes, biopower is linked to hegemony as opposed to the much narrower concept of state policy. In Potchefstroom 'letting die', or ensuring a minimal level of existence on the outside to secure the inside, has some obvious function to it. By letting die the root causes of crime are fuelled and this ironically gives legitimacy to laagerisation and narrowly defined spaces of representation. At the same time criminality is afforded a space or set of spaces, under the radar, so to speak. Contemporary hegemony needs this type of abandonment, although I am not arguing for what Li (2010:66) critiques as a 'stealthy functional equilibrium' whereby some need to die so that others can live. My point is that stigma has had the effect, as far as crime is concerned, of providing justifications for laagerisation that aids abandonment. I would agree with Howell and Richter-Montpetit (2019) that race also largely *constitutes* the particular biopolitics observed in the study area. It is indeed sometimes gratuitous, as the introduction to Part III will demonstrate. For the most part, however, subjects, where wittingly engaging in violence, (erroneously) consider it instrumental. It is also not an all-encompassing onslaught on those deemed superfluous, as Mbembe (2003) articulates Necropolitics in the case of Israel and Palestine.

As regards the application of Wacquant's notion of advanced marginality, this chapter has shown that increased policing of stigmatised spaces might not fit this context precisely. Limited policing in some of these areas is so significant that mob justice has increasingly become common place (Anon, 2019:5, 10 November 2019; Saayman, 2019:3, 5 September; Leshage, 2018b:2, 26 July). However, similar injuries of stigmatisation have been observed, through the isolation of and disinvestment from, stigmatised spaces. At the same time, elites have demonstrated the ability to react to and largely arrest stigmatisation, where it matters to them.

Given the dynamic nature of laagerisation and the increase in technologies by which border-making takes place, the dialectic between representations of space and spaces of representations in Potchefstroom is likely to remain critical. In the future, further and deeper displacement of crime onto other segments of the city seems likely. The directions of such displacement might be significantly influenced by what is allowed for through the execution of official spatial development policy and how its execution is mediated through interaction with the middleclass and their political representatives. There can be little doubt that crime discourse will directly and indirectly have a bearing on these developments. As far as floating signifiers are concerned, the overwhelming problem in need of action is the conflation of particular spaces and population groups with crime. These need to be publicly challenged and replaced by a more productive and inclusive set of articulations that acknowledge the entanglements that have remained functional despite laagerisation and that address the legal and ethical characteristics of these enduring entanglements. While Chapters 1–3 have established a more macro-level perspective on the broad strokes of security infrastructures and the politics of crime South Africa and Potchefstroom the following section, Part II, will deal with more specific infrastructures individually, mostly focussed on the ideational flows the facilitate.

References

Adesina, J. 2019. Globalization, migration and the plight of Nigerians in South Africa, in O. Tella (ed.) *Nigeria-South Africa relations and regional hegemonic*. New York: Springer, pp. 109–127.

Anon. 2019. Infuriated community members torture suspects over stolen goods. *Potchefstroom Herald*. 10 November, p. 5.

Anon. 2018. Geen geld vir veiligheid. *Potchefstroom Herald*. 3 May, p. 4.

Anon. 2017. SA has 6th most shopping centres in the world and building more in Fin24. 24 December. Online: www.fin24.com/Companies/Retail/sa-has-6th-most-shopping-centres-in-the-world-and-building-more-20171224. Date of Access: 28 June 2020.

Arendt, H. 1953. Ideology and terror: A novel form of government. *The Review of Politics*, 15(3), pp. 303–327.

Bangani, Z. 2019. Domestic workers: Overworked, underpaid and unprotected. *Mail & Guardian*. 1 July. Online: https://mg.co.za/article/2019-07-01-domestic-workers-overworked-underpaid-and-unprotected/ Date of access: 24 July 2020.

Cheteni, P., Mah, G. and Yohane, Y.K. 2018. Drug-related crime and poverty in South Africa. *Cogent Economics & Finance*, 6(1), pp. 1–16.

Cummins, I. 2016. Wacquant, urban marginality, territorial stigmatization and social work. *Social Work*, 28(2), pp. 75–83.

De Villiers, B. 2020. Opinion: Domestic workers are undervalued, underpaid. *Dispatch-Live*. 18 March. Online: www.dispatchlive.co.za/news/opinion/2020-03-18-opinion-domestic-workers-are-undervalued-underpaid/ Date of access: 24 July 2020.

Dlamini, K. 2019. Fighting corruption essential to talking the heroin trade in SA. Online: www.corruptionwatch.org.za/fighting-corruption-essential-to-tackling-heroin-trade-in-sa/ Date of access: 8 April 2020.

Foucault, M. 2003. *Society must be defended*. New York: Picador.

Harris, B. 2002. Xenophobia: A new pathology for a new South Africa? in D. Hook and G. Eagle (eds.) *Psychopathology and social prejudice*. Cape Town: University of Cape Town Press. https://www.files.ethz.ch/isn/100580/xenophobia.pdf.

Howell, A. and Richter-Montpetit, M. 2019. Racism in Foucauldian security studies: Biopolitics, liberal war, and the whitewashing of colonial and racial violence. *International Political Sociology*, 13(1), pp. 2–19.

Isike, C. and Isike, E. 2012. A Socio-cultural analysis of African immigration to South Africa. *Alternation*, 19(1), pp. 93–116.

Jansen van Rensburg, N.S. 2006. The first 'white' town north of the Vaal: Inequality and apartheid in Potchefstroom. *New Contree*, 51, pp. 131–147.

Keppler, V. 2019. 'n Daggawolk hang oor Potchefstroom. *Vrye Weekblad*. 21 June. Online: www.vryeweekblad.com/nuus-en-politiek/2019-06-21-n-daggawolk-hang-oor-potchef-stroom/ Date of access: 28 April 2020.

Laclau, E. 1990. *New reflections on the revolution of our time*. London: Verso.

Lefebvre, H. 1991. *The production of space*. Translated by Donald Nicholson-Smith. London: Blackwell.

Leshage, S. 2018a. ANC whip says Vyfhoek development will go ahead. *Potchefstroom Herald*. 3 May, p. 4.

Leshage, S. 2018b. Man dies after mob justice attack. *Potchefstroom Herald*. 26 July, p. 2.

Li, T.M. 2010. To make live or let die? Rural dispossession and the protection of surplus populations. *Antipode: A Radical Journal of Geography*, 41(1), pp. 66–93.

Matsinhe, D. 2011. Africa's fear of itself: The ideology of Makwerekwere in South Africa. *Third World Quarterly*, 32(2), pp. 295–313.

Mbembe, A. 2003. Necropolitics. Translated by Libby Meintjes. *Public Culture*, 15(1), pp. 11–40.

Morris, A. 1998. 'Our fellow Africans make our lives hell': The lives of Congolese and Nigerians living in Johannesburg. *Ethnic and Racial Studies*, 21(6), pp. 1116–1136.

Neocosmos, M. 2008. The politics of fear and the fear of politics: Reflections on xenophobic violence in South Africa. *Journal of Asian and African Studies*, 43(6), pp. 586–594.

Park, Y.J. and Rugunanan, P. 2009. Visible and vulnerable: Asian migrant communities in South Africa. Centre for Sociological Research. University of Johannesburg. Auckland Park.

Pillay, N.K. 2010. An exploratory study of the citizenship process of immigrants to South Africa: A case study of Pakistanis in Durban. Mini-dissertation for the degree Master of Population Studies. University of Kwazulu-Natal: Durban.

Pinnock, D. 2019. Criminalisation key to Cape Town's drug wars. Online: www.news.uct.ac.za/article/-2019-08-30-criminalisation-key-to-cape-towns-drug-wars. Date of access: 8 April 2020.

Povinelli, E. 2011. *Economies of abandonment: Social belonging and endurance in late liberalism*. Durham, NC: Duke University Press.

Rose, N. and Miller, P. 1992. Political power beyond the state: Problematics of government. *The British Journal of Sociology*, 43(2), pp. 173–205.

Saayman, M. 2019. Klopjag lei na geweld. *Potchefstroom Herald*. 5 September, p. 3.

Shaw, M. 2002. West African criminal networks in South and southern Africa. *African Affairs*, 101(404), pp. 291–316.

Shearing, C. and Stenning, P. 1981. Modern private security: Its growth and implications, in M. Tonry and N. Morris (eds.) *Crime and justice: An annual review of research*. Chicago: University of Chicago Press, pp. 193–246.

Sicetsha, C. 2019. Narcos report ranks South Africa as largest market for illicit drugs in sub-Sahara Africa. *The South African*. Online: www.thesouthafrican.com/news/narcos-report-south-africa-largest-market-illicit-drugs-sub-sahara-africa/ Date of access: 8 April 2020.

Slater, T. 2015. Territorial stigmatization: Symbolic defamation and the contemporary metropolis, in J. Hannigan and G. Richards (eds.) *The handbook of new urban studies*. London: Sage, pp. 111–125.

Taylor, D. (ed.) 2011. *Michel Foucault: Key concepts*. Durham, NC: Acumen.

Thompson, D.K. and Grant, R. 2015. Enclaves on edge: Strategy and tactics in immigrant business spaces of Johannesburg. *Urban Forum*, 26(3), pp. 243–262.

Van Riet, G. 2020. Intermediating between conflict and security: Private security companies as infrastructures of security in post-apartheid South Africa. *Politikon: The South African Journal of Political Studies*, 47(1), pp. 81–98.

Wacquant, L. 2016. Revisiting territories of relegation: Class, ethnicity and state in the making of advanced marginality. *Urban Studies*, 53(6), pp. 1077–1088.

Wacquant, L. 2008. *Urban outcasts: A comparative sociology of advanced marginality*. Cambridge: Polity Press.

Waquant, L. 2007. Territorial stigmatization in the age of advanced marginality. *Thesis Eleven*, 91(1), pp. 66–77.

Wacquant, L., Slater, T. and Pereira, V.B. 2014. Territorial stigmatization in action. *Environment and Planning*, 46, pp. 1270–1280.

Wetdewich, D. 2018. Rowers skiet student vir niks. *Potchefstroom Herald*. 12 April, p. 3.

Wetdewich, D. and Saayman, M. 2018. Hulpkreet val op dowe ore. *Potchefstroom Herald*. 19 April, p. 3.

Wolpe, H. 1972. Capitalism and cheap Labour-power in South Africa: From segregation to apartheid. *Economy and Society*, 1(4), pp. 425–456.

Cited interview

Estate agent. Interviewed by honours research methodology students. 20 April 2018. Potchefstroom.

Part II

The echo chamber effect

If you look at the newspaper and if you monitor Facebook and so forth, it really sounds to me as if things are on the decline on the Bult and in the CBD. I see stories of people who are assaulted. I really avoid those areas, even if I want to buy something there. I won't go to the Bult in the evening. And people are scared. The Bult is close by and all those flats that are being built; all those nodes are expanding towards the residential (suburban) area. It is a different type of resident. Some of the crime can expand towards us. I am afraid of more violent crime that can shift this way. It is already happening in Van der Hoff Park and Baillie Park. With the river here it is easier for criminals to escape (Interview with Oewersig resident, November 2018, author's translation).

*****

Fear is not something to be judged. One cannot employ a constructivist framework and pass judgement on what would be an appropriate level of fear. This interviewee is afraid and based on the reasons she gives, it is understandable. She reads the local newspaper. She follows social media and as an informed resident she has developed a view of the status quo. Further qualifications are required. As the comparative statistics in the introductory chapter suggest, Potchefstroom does, seemingly, have a significant amount of crime, also relative to other contexts. This crime is largely property related and not as violent as it is in many other parts of the country. Still, violent crime does happen and fearing something because it is a real possibility, is not irrational. Such fear might even be healthy. But, fear can also become a divisive force, especially when it is not well contextualised. Levels and especially forms of crime do not correspond perfectly with what the discourses that circulate through social and conventional media and other public platforms might suggest. Moreover, fear of a mystified criminal indicates a lack of such contextualisation. The situation is worsened when divisive discourses, through daily circulation, are attached to the reality of crime. There is a need to demystify criminality and to disentangle problematic discourses from a healthy vigilance in response to crime. We can feasibly do this through a constructivist framework, such as the one employed in this book. Contextualising crime does not mean stating the 'optimal' or 'correct' way of viewing the matter, as that would simply be arrogant and an impossible fool's errand. Instead alternative perspectives may alleviate some fear, while potentiating meaningful social

DOI: 10.4324/9781003028185-6

praxes. By praxes I mean intervention into the intertwined problems of crime and an evident lack of social cohesion.

Much legitimate knowledge of crime passes through security infrastructures in order to help citizens protect themselves. In addition, however, security infrastructures such as CPFs, newspapers and the social media sites of PSCs also (in)advertently perpetuate problematic ideas associated with crime. These ideas include floating signifiers and sutured identities that may be challenged, as they perpetuate discord. This section, Part II, develops the notion of an echo chamber effect through which problematic ideas ancillary to the reality of crime are amplified. The focus here is primarily on ideas that, following a Laclauian logic, is only one part of discourse, which also includes practice. Discursive formations are often derived from the physical and ontological structures discussed in Part I. They are also reaffirmed through some of the routine security practices that are the focus of Part III.

This section is divided into three chapters. The first deals with the CPFs. The Potchefstroom precinct is divided into four CPF sectors, each with its own forum, which convenes monthly. These forums are intended as spaces where the public and the police interact and share concerns and information on crime. It is also a structure through which the police can be held accountable. Sub-sectors have officially sanctioned neighbourhood watches. Community policing is governed through an overarching management structure at the precinct level, The CPF, which includes representatives from all four sectors. The CPF has a vibrant social media presence, most notably through its Facebook page, JB Marks Security and subsector or neighbourhood WhatsApp groups. There is also an overarching WhatsApp group for all group coordinators. The analysis points to the fragility and dependence of the CPF on particular motivated individuals to function. Given enough time, it seems, the collapse of sub-forums is a given. Chapter 5 deals with the ideas shared through the local newspaper, the *Potchefstroom Herald*. This analysis reveals a significant reliance by this publication on crime to fill its pages. Just like the social media analysed in Part II, this publication sometimes also too serves as a platform for problematic discourses, ancillary to the problem of crime. Chapter 6 scrutinises the public Facebook pages of PSCs to explain how these platforms securitise security infrastructures, including the laager in general. As with the CPF Facebook page and the *Herald* the public also adds content to these platforms.

Overall, Part II demonstrates how particular ideas circulate through three different sets of infrastructure. In the process unfavourable and divisive ideas in the service of frontier governance are attached to crime. The result is that exclusionary spaces of representation are often reinforced. These ancillary discourses may productively be conceived of as floating signifiers that may be challenged to incrementally address the social division associated with crime. At the same time, due care is required to combat crime and social divisions. The overarching argument of this book is that these two imperatives must as far as possible be approached as mutually reinforcing and not mutually exclusive.

4 The community policing forum

The talk shop and the circulator of ideas

Introduction

The first section of this book outlined a particular conception of the physical and ontological milieu Potchefstroom residents typically occupy. As a collective infrastructure, the laager through its constitutive parts shape the flow of ideas and practices. This chapter is the first of three chapters that deal with the circulation of ideas, more so than practices. The current chapter does this through an analysis of the CPF as a security infrastructure. This chapter is based on my notebooks, especially the notes I took while attending Sector One and Sector Two' monthly meetings during 2018 and 2019. I also attended one meeting of the Sector Four CPF, formed during 2019, following the division of Sector One. I did not attend meetings for Sector Three which includes farms that surround the city and thus fall outside of the remit of the current study. In addition, the chapter draws on the circulated minutes of meetings from late 2017 until the end of 2019 for Sector One and key informant interviews with CPF members who have a long institutional memory while working. I also accompanied a Sector Two CPF neighbourhood watch on two occasions. Finally, the chapter analyses the discourse which circulates through the Potchefstroom CPF Facebook page, renamed JB Marks Security subsequent to my data gathering.

The chapter starts with a literature review on community-oriented policing (COP) globally. This section concludes by briefly sketching the context of its adoption and its successes and shortcomings in South Africa. The chapter then engages with the case of Potchefstroom. It firstly analyses the functioning and periodic decline of CPF structures, as well as the discourses that circulate in CPF meetings. Social media has been a major tool of COP in Potchefstroom, with numerous successes and shortcomings. These are assessed in the subsequent section. The final section of this chapter picks up on the observed thread of apathy amongst the public. Sporadic interest, interspersed with, at times, vocal criticism of the SAPS, has not allowed the CPF to provide an entirely useful bridge between the SAPS and society. There is, however, a small yet critical mass of interested parties willing to engage with the SAPS in good faith. It is imperative that these voices supersede the problematic discourses and signifiers espoused by other residents.

DOI: 10.4324/9781003028185-7

Community policing forums globally

COP emerged in the USA in the 1970s and gradually spread around the world. Today practices called COP are found throughout the world (Nalla and Newman, 2013). COP represented a philosophical shift in policing practices, albeit that the concept remains rather ambiguous (Demirkol and Nalla, 2019:692). The move to COP was partly brought on by the results of existing policing practices in the USA, which caused many academics and police executives to call for a change (Reisig and Giacomazzi, 1998:548). Today, there is very little agreement on the exact details of this form of policing across different contexts. But, there are more typical practices and assumptions associated with COP that can be cited here. The practice of COP is often linked to more diffused areas of operation, collaboration between the police and the public and a greater focus on prevention. This form of policing manifests at different levels. At the individual level, a citizen may report suspicious behaviour. At the level of the collective members of the community may participate in various programmes and forums (Reisig and Giacomazzi, 1998:548). In general, neighbourhoods tend to become the units of analysis (Cordner, 2010:444). Even with a trimmed down state, which transfers greater responsibility to the public at lower levels, there are still particular functions that the state police are meant to provide. Community policing in some parts of the world does not change the fact of visible policing and police officers walking their beat. It simply changes which beats are prioritised, based on information shared by the public and seemingly analysed by the police. The police would logically report back to the community on the state of crime levels and interventions that followed from the initial reporting of crime to the police. This requires collaboration between the public and the police and not just the criminal justice system, narrowly defined. COP is meant to be based on problem solving (Moore, 1992). Moreover, it requires a type of bidirectional feedback between the police and the community that is meant to build trust and community networks (Boettke et al., 2016:318; Cordner, 2010:442). In contexts where there is a lot of crime, however, Morabito (2010:570) cites various studies that point to an inability by the police to dedicate sufficient resources to community policing.

Community policing is therefore meant to emphasise prevention and it is evidence based, as the information shared through interaction between the police and the community should help set priorities for intervention. As such, partnerships between the police and the public can take on various forms, such as neighbourhood watches, neighbours watching each other's homes and reporting local crime, such as drug dealing. COP can be part of problem identification and problem solving. This may be through policing broadly defined or educational and recreational activities. Other elements that might be included in community policing are increased access of the public to police officers through bicycle and foot patrols, the use of satellite police stations and the long-term assignment of officers to specific beats (Reisig and Giacomazzi, 1998:587). Decentralised 'mini-police stations' seem to be appreciated by the communities of the global

North (Cordner, 2010:447). There is also a recent example of this in Potchef-stroom, to be elaborated on in Chapter 9. The question can however be asked: who is it that appreciates these stations? And, are these stations merely an infra-structure of existing power relations, privilege and hegemonic orders in society?

In response, proponents of COP would emphasise that this approach to polic-ing is intent on addressing the causes as well as the symptoms of crime (Cord-ner, 2010:441) and that there is a focus on social cohesion as opposed to simply removing lawbreakers from society. In a sense, a greater emphasis on under-lying causes is required, which traditional policing had not resolved (Samp-son and Raudenbush, 1999). For example, COP may entail police run youth programmes (Cordner, 2010:440; Reisig and Giacomazzi, 1998:547; Maguire et al., 1997).

This approach to policing, though arguably necessary in some or other form, might be a far more problematic undertaking than proponents would like to acknowledge. Many, especially macro-level, root causes explained in Chapter 1, might not fall within the ambit of the local community, let alone the local police station to address. In the South African case, large-scale disquiet, based on ineq-uities in educational and job opportunities are surely examples. Such inequities might be influenced by how the government has struggled to position itself favour-ably within the world economy and the large-scale looting of state resources that has been a preoccupation of perpetrators and their adversaries in government for over a decade.

According to Cordner (2010:443) COP requires some restructuring at the level of the police station. Decentralisation is one such strategy, whereby the police sometimes delegates responsibility, so that some police officials act more inde-pendently. Restructuring might also require 'flattening', or making the police command structures less hierarchical. The rationale here is to reduce rigidity, bureaucracy and to improve communication. Restructuring might also require de-specialisation within the police in order to offer more direct services to the community. The police might also create teams to encourage more effective and quality solutions to problems being developed.

How effective has community policing been elsewhere? The evidence seems to vary. A slight majority of studies have shown a decrease in crime levels. However, as Cordner (2010:445) notes, variations in research designs limit the authority of such a conclusion. Moreover, these studies have typically been con-ducted in northern contexts. Unlike South Africa, northern police stations tend to be better resourced and as such have more police officers to dedicate to com-munity policing. There also tends to be more trust in the police in the North, notwithstanding the problematic nature of policing, brought to light in the case of the USA, once more with the murder of George Floyd in 2020. There have also been thought provoking publications by authors such as Vitale (2017) who emphasises the inherent brutality of the police, which requires defunding this organisation and redistributing funds to social projects. Apart from these recent examples, the public in general, seem likely more willing to cooperate and even collaborate with the police in many northern contexts. There is evidence of a decrease in the fear of crime in these contexts (Demirkol and Nalla, 2019:693: Boettke et al., 2016:313; Cordner, 2010:445). Still when this perception of decreased crime and trust in the police fails, the programme of COP fails. Thus COP was abandoned in New York, which according to Boettke et al. (2016:313) was once considered the flagship COP programme in the USA. Moreover, where there is a question about the durability of the project, community members tend to lose interest and seize to participate (Boettke et al., 2016:319). As the analysis in this chapter will show, COP in Potchefstroom has historically been hamstrung by a lack of interest. At the same time residents display a severe lack of trust in the SAPS. Further evidence of the largely mixed results in research on COP is found in Gill et al's (2014) meta-analysis of 25 reports. They find that COP has a positive effect on citizen satisfaction and perceptions of disorder and police legitimacy. Conversely, they find that COP has a limited impact on crime and fear of crime. As we will see in the following, apathy and distrust in the police makes people far less willing to collaborate with this organisation through community policing. Furthermore, if we consider the evidence cited from social and other media in this chapter and in Chapters 5 and 6, there seems to be immense fear of crime in Potchefstroom, regardless of community policing.

Another important variable in determining the success of COP is the institutional culture of the police. Cordner (2010:447) in summarising the state of the literature on COP argues that police officers often resist community policing as a new strategy, based on philosophical disagreements, doubts over the efficacy of COP or simply because of habit. Demirkol and Nalla (2019:693) come to similar conclusions. As they put it, others have found that, despite an increase in budgets for COP, it was often 'business as usual' for many police stations. A more traditional policing culture that prioritises ruthless response does not bode well for COP. Demirkol and Nalla (2019:702) in their research measuring the influence of police culture on COP in a Turkish case study find that police culture indeed limits the willingness of police officers to comply with the call for COP. This does however vary within police departments. Some officers are more enthusiastic than others when it comes to COP, as attitudes, norms and values differ. Demirkol and Nalla (2019) essentially, following Palmiotto et al. (2000), recommend changes in how recruits are trained. This should include social history, teaching the community as a living organism and more COP specific skills and knowledge. Institutional culture within the police and apathy from the public has often translated into police-community interaction merely for the sake of having such interaction (cf. Rosenbaum and Lurigio, 1994). Similarly, Moribato (2010:568) reports on various studies that indicate how COP has been implemented in very superficial and oversimplified ways. Although there is some controversy over this, attitudes towards the police might affect community participation in COP. Furthermore, older residents, who are more security conscious might be more interested in collaboration, while younger residents often view the police as a constricting force when it comes to their freedom (cf. Reisig and Giacomazzi, 1998:549–550, 556).

Although once labelled as the dominant paradigm on policing, COP has for the various reasons cited earlier been subject to much criticism in recent years. The following discussions will confirm, be it with a local inflection, many of the challenges for effective COP.

Community policing in South Africa and Potchefstroom

In South Africa COP emerged in the early to mid-1990s as a response to the poor image of the SAP, as an oppressive tool of the apartheid state. Community policing would become a means whereby the perceived distance between the police (now a 'service' and not a' force', hence SAPS) and society would be closed through frequent discussion on community crime concerns. Community policing platforms is mandated by the *SAPS Act 68 of 1995* (South Africa, 1995) and the *Interim Regulations on Community Police Forums and Boards* (South Africa, 2001). Although laudable, the purpose of bringing state and society closer together on the matter of crime has not always been achieved. Today South Africa is in crisis (Van Riet, 2020). The country is struggling with crippling debt. The Covid-19 pandemic proved to be the final straw that led Moody's to adjust South Africa's sovereign credit rating to below investment grade. The country is now rated as such by all three major credit ratings agencies, also include Standard and Poor's and Fitch. The country has experienced considerable corruption. There is a recurring energy provision crisis and there is a clear lack of social cohesion, for example evidenced in continued and incessant racism (Van Riet, 2016) and racial discourse. Vast unemployment, coupled with passive income possibilities for a very small minority has made it one of or even the most unequal society in the world. It could be argued that levels of trust between South Africans are the lowest it has been since the end of apartheid. Trust in institutions such as the SAPS is one symptom of this, although I do not want to give the impression that trust in the police ever recovered since the end of apartheid. It seems more likely that a greater spread of South Africans now distrusts the police, be it for different reasons. Hence today still, a 'gap' between the police and society exists. The SAPS are distrusted, sometimes for good reasons, although sometimes they are possibly judged unfairly. So, in theory, the need for effective COP in South Africa is clear. The practice has, however, often struggled to gain and maintain traction.

Due to greater concerns with crime around the North-West University and the sheer size of Sector One, which included virtually all of the retail spaces in Potchefstroom, this sector was divided and Sector Four was formed in 2019. Sector Four covers the most northern parts of Potchefstroom, while Sector One was reduced to cover most of central Potchefstroom, North of the N12 highway between Johannesburg and Cape Town. Sector Two covers the parts of the city to the South of the N12. As mentioned, Sector Three includes the rural areas around Potchefstroom, which are not the focus of this book. Map 4.1 provides a visual representation of the CPF sectors for the Potchefstroom police precinct.

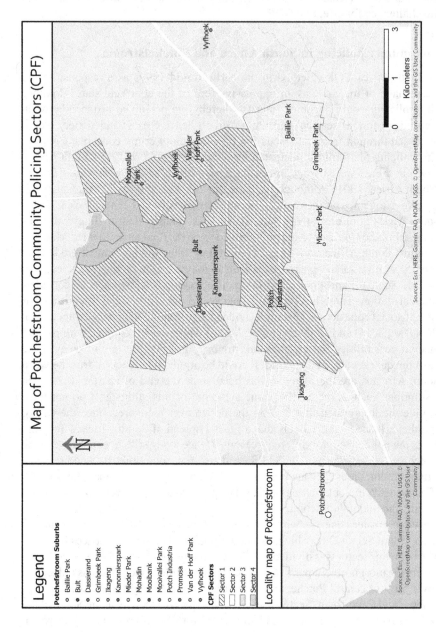

Map 4.1 The community policing forum sectors for the Potchefstroom police precinct

According to a former policeman, with experience beyond Potchefstroom and a longstanding CPF member in Potchefstroom, these structures have, as some within the international literature report, often been mere compliance exercises. Merely having meetings and conducting some arbitrary activities would serve as evidence of legislative compliance on the part of the SAPS. At its worst, the public would use these structures for other means. In historically white areas, these forums would often serve as platforms for complaints over municipal service delivery, where the services in question sometimes barely, if at all, relate to crime. In historically black precincts CPFs would become platforms for would-be politicians to make a name for themselves. Indeed, CPF members in Ikageng often become municipal councillors (Interview with former policeman, July 2017; Interview with CPF office bearer, February 2020).

There have also been discrepancies between the literature from the global North and the experience I have had researching this particular part of the global South. Increased foot patrols for example have been a common feature in northern contexts. In a context such as South Africa, where the police to populace ratio is so small, this, unfortunately, is not a practical solution. Neighbourhood watches officially mandated through the CPF are sporadic. Since the end of 2019 remote viewing via CCTV cameras by PSCs appear to have replaced some neighbourhood watches in more affluent parts of the city. This is a fast-developing trend, as it rapidly spreads from North to South through Potchefstroom. Such remote viewing might be useful. But, CCTV cameras monitored by PSCs arguably carry with them the potential for intrusive practices and certainly a greater gap between the SAPS and certain segments of population, as the SAPS are increasingly bypassed by wealthier residents who rely on PSCs for security.

The CPF has also been shedding members to the Afriforum neighbourhood watches. It seems that those leaving, view the CPF as too timid and too closely related to a perceived dysfunctional state. The Afriforum neighbourhood watches however, intensions aside, have the potential for further divisiveness. Afriforum is a civil society organisation associated with defending very narrow white interests, although it should be noted that this portrayal has been vehemently denied. Still, this perception along with some evidence of how their neighbourhood watches operate, and the types of CPF members that have gravitated towards Afriforum is a potential concern. For example, on one evening I climbed into the back of a neighbourhood watch vehicle. Next to me was a bag with cable ties sticking out. Cable ties make sense in the unlikely event of a citizens' arrest. However, the owner and driver of the car struggled to explain away the axe, which was also partially sticking out of the bag. The driver of this car and the other CPF neighbourhood watch member in the care that night informed me that they were soon ending their relationship with the CPF and joining the Afriforum neighbourhood watch. Furthermore, in one CPF meeting the police station commander reported that there was an incident where an Afriforum neighbourhood watch vehicle aggressively forced an unmarked police vehicle off the road. Such behaviour is not within the ambit of a neighbourhood watch. Their job is to observe and report and to only intervene when they are certain that the specific offences, described in Schedule I of the *Criminal Procedure Act* (South Africa, 1977) are about to or have been committed.

Legislative compliance: the limited circulation of useful ideas

As a legislative requirement in South Africa, I often got the sense that community policing was done begrudgingly and superficially. As mentioned earlier, the Potchefstroom precinct CPF is divided into four sub-forums. Each of these forums meet on a monthly basis. Each meeting, based on those I have attended and those of which I have documentation/minutes, cover a relatively standard set of issues and a relatively standard agenda. Standing items include the following. The SAPS officer designated to the sector would confidentially share the most prevalent crimes in the area, based on reported cases. This information is then shared through social media, as far as it is legal to do so. Some crime statistics and facts are embargoed and cannot be shared with the public until the embargo is lifted after the national police commissioner has signed off on them. This is mainly to prevent tipping-off criminals about the information the SAPS possess. Crime information shared in the sub-forum meetings would relate to prevalent crimes for the previous month, where they occur, on which days and at what times they typically occur. Then there are opportunities for members representing smaller segments of the sector, such as suburbs, to raise matters of concern pertaining to their area. Finally, the meeting discusses, often very briefly, crime prevention through environmental design, as a standing item. This further confirms the understandable though not necessarily useful in the long run, approach that the SAPS are – I believe forced – to take in the matter of crime. Technical solutions are sought that deal with crime at the level of the individual criminal, without very many interventions demonstrating an acknowledgment that crime is also a societal problem. In other words, the social programmes or other more deeply preventative measures noted in the international literature discussed earlier, does not feature in CPF meeting or the disposition of the SAPS articulated in these meetings.

Minutes of meetings often have very little substance and there are doubts amongst interviewees and the public, judging from social media content, that concerns raised in the meetings are indeed addressed. Reporting back on these matters have at best been limited or inadequate. According to members who have extensive institutional memory the reporting back on issues, beyond those generally raised by the public through the CPF, remains weak. This is something the SAPS must take heed of. For people to feel safe, they need to understand what is being done about their concerns. Such reporting back along with the proper functioning of structures such as neighbourhood watches informed by constitutionalism, include rights such as freedom of movement, might limit the laagerisation of security provision in the city. As such, proper reporting back to the population might better serve social cohesion than the likes of custom-built solutions by each neighbourhood or CIDs, which are the topics of later chapters in this book. Not by any means is all of this the SAPS' fault. How can street lights for example, be kept on to render a neighbourhood safer at night when the municipality is basically bankrupt? One might think that such service delivery issues are forwarded to the relevant authorities. But what is there to report back on, when these authorities have their hands tied financially?

As it stands, these meetings seem more like a compliance exercises than actual substantial community policing. The CPF is most significant as a structure which circulates ideas, which impresses upon the public an individualist and environmental design focus on crime prevention and response. There are limits, given their resource constraints, to the extent of blame we may apportion to the SAPS. The literature cited earlier does state that community policing is more successful in larger organisations, as opposed to under capacitated ones.

It does seem that since the election of the new CPF management in 2019, that meetings in Sector One did become a bit more substantive with clear actionable decisions being made and recorded in minutes. These minutes identify a person responsible for executing each decision. It would appear that more substantive feedback from each neighbourhood, leads to more meaningful discussion and concrete action. Hence there are signs of evidence-based policing (EBP) emerging, where meetings are not rushed through, but used as a platform arguably more within the spirit of COP. This is rather encouraging, provided the pattern continues and importantly, provided that this productive trend grows. It is still too often the case that representatives from neighbourhoods do not avail themselves for these meetings and as such forfeit their opportunity to meaningful COP. The CPF is also a place where different neighbourhoods learn from each other. Through these structure ideas and practices around, for example, live monitored CCTV cameras circulate. It is sometimes recorded in the minutes of meetings that a representative of one neighbourhood will assist another neighbourhood with particular tasks, such as acquiring technology and associated services.

With the periodic collapse of sector forums to the newly found impetus in one sector, it does appear that successful COP often depends on the energy and stamina of key individuals to drive the process, both on the side of the community and the police. Luckily, there are such individuals who have had both the energy and the stamina throughout the rise and decline of sub-forums. Some of these individuals have created more enduring social media platforms in the name of the CPF. Unfortunately, along with disseminating useful safety tips, these platforms also facilitate the circulation of potentially divisive ideas attached to crime.

Social media: circulation, favourable and less favourable

For a number of years, the CPF have benefitted from a limited number of very committed members. Their development and management of social media platforms has demonstrated this most clearly. There is a JB Marks Security Facebook group (known as the Potchefstroom CPF group while I was doing my research) and there are WhatsApp groups for sub-forums and sometimes even smaller units, that is neighbourhoods. A lot of information is shared, by CPF members and the public more generally on the Facebook group especially. I was a member of the Sector One WhatsApp group for a while, but not a lot of information was shared through this platform. There is also a coordinators' WhatsApp group, where the administrators of all the smaller groups decide what content will be shared with which groups, based on information received from the SAPS and the public. Having such a decision-making structure is possibly wise on the part of administrators. Generally, they only share information they deem to be immediate threats. Limiting commentary that garners of fear is arguably more easily achievable on WhatsApp than Facebook, as these groups have less members. Although some less useful information has been shared on WhatsApp, moderators have been able to, because of the smaller audience, shut down a particular discussion upon careful interrogation of the information. Facebook on the other hand includes the information shared on WhatsApp plus additional information shared by the SAPS and notably the public more broadly. The CPF often shares credible and useful information on both platforms and as such, despite the potential these sites create for the circulation of problematic ideas, the enduring use of social media has arguably been the CPF's most significant achievement. This section will present the good along with the bad. It will show how social media has been useful and how it has served the circulation of potentially problematic and divisive ideas. The discussion here cannot be exhaustive. It merely serves as an illustration of the types of ideas that circulate through CPF social media platforms. I will also make suggestions as to how these might be adjusted in line with democratic principles, without sacrificing the good that these platforms may serve in helping people feel safe or even to help reduce crime. Regarding the latter, caution is however required. As will be argued in subsequent chapters, to add security infrastructures in one part of the city or even the entire city, might not necessarily reduce crime as much as it displaces crime elsewhere. To believe differently, is rather shortsighted, as these technologies cannot address many of the root causes of crime identified previously. They cannot eliminate the motives for crime. Still, we cannot blame people for taking action to feel safer. The question we do need to ask is, at what point do these technologies move beyond serving a security function and become a means of harassment and start to serve exclusionary practices? As with the argument throughout this book, this discussion also demonstrates that Potchefstroom's citizenry, just like that of any other territory, includes a variety of individuals. This is a source of hope, as it potentiates dynamic and productive collaborations.

A lot of information is shared in the form of crime trends and safety tips. This content might include warnings of remote jamming devises used in parts of the city. These devises interfere with the central locking mechanism of a car. The group administrators might also share information on recently stolen vehicles, with plate numbers, and ask the community to contact the SAPS should they see those vehicles. Other public service type announcements include reports of missing persons or the date, venue and times of CPF sector meetings.

Especially since the new station commander took over in 2018, the SAPS have sometimes used the CPF Facebook page to report back on successful apprehensions. I have observed an increase in the social media presence of senior SAPS officials and other members of the criminal justice system since 2019. These reports elicit positive and negative responses. For the purposes of my argument, the fact that there are positive responses is significant and encouraging. In due course the negative responses might become isolated and such feedback that has been missing in the past, might become a useful tool by which to build trust between state and society as far as crime reduction is concerned.

On the negative side, the CPF's social media platforms have, likely unwittingly for most part, fuelled racial divisions, for example, in the way descriptions of suspects are given. Sometimes this is unavoidable. Such descriptions may, however, perpetuate an association of black people with crime, without nuance. Descriptions of suspects or arrested perpetrators are typically devoid of proper contextualisation of the relationship between race, class and crime. Once again crime, unlike the explanation I provided in Chapter 1, becomes an attribute of the perpetrator only, and not also of the social order which contributes to the shaping of crime patterns and perpetrators. It is under these conditions of limited contextualisation and a social order where dominant groups lack sufficient self-awareness, that race becomes a symbol of the abject. Consider the following quotation.

> Good afternoon people. I would like to bring it under your attention that there are once more people who are walking and 'raising funds'. Just now there was a black man in our street with a crumpled piece of paper and he smelled of pure alcohol.
>
> (Author's translation, JB Marks Security
> Facebook group 6 October 2017)

The membership of the JB Marks Security Facebook group is diverse. The content posted by the public is, however, dominated by white and Afrikaans people. This has often had the adverse consequence of foregrounding particular narratives, which although not explicitly framed as such, might be described as reflecting notions of an 'onslaught' on a narrowly defined 'cultural group'. As mentioned in Chapter 2, just like the total onslaught engineered by the apartheid regime to justify extraordinary measures in response to a ubiquitous 'communist threat', a new seemingly black, criminal threat is articulated by some community members. But is a difference. The total onslaught of the 1980s was a creature created by the state to justify itself. Yet, as one peer reviewer of my work rightly remarked, nearly every South African has a crime story. This is true. South Africa has a lot of crime. This crime is often violent, although in the case of Potchefstroom violent crime does not dominate. The notion of onslaught however comes in, exactly where a new narrative of persecution is attached to the criminality that we do indeed face. In other words, this notion of onslaught or persecution, is not inherent to the problem of crime. Instead it is an add on that potentially fuels division and there is no more emotive issue in this regard than farm murders. The murder of farmers has been part and parcel of South African public discourse for decades. It is often taken as a political vendetta against the Afrikaner. Yet, discussions with a senior white, Afrikaans SAPS officer (2018) reaffirmed a widely held reading of the matter, namely that these are, *for the most part*, crimes of opportunity (cf. Clark and Minnaar, 2018). They are the consequence of housebreaking or house robbery that escalates, while the initial targeting was based on the remoteness of the target. Consider the following example.

> Last night's farm attack – again Rysmierbult. The dogs barked. *Oom*[1] went outside at around 21:15. *Skelm*[2] jumped over the wall and placed a gun against his (the *Oom*'s) head. He took him into the house. The wife heard that the Oom was talking to someone. It did not sound right and she fetched the shot-gun. [The] other one, armed with [a] knife stabbed the *oom* in the hand. The *oom* ran to the [bed]room and the robber fired two shots towards him. Fortunately, he missed. The *oom* got a gun and also fired two shots. The robbers fled and they left with two cell phones. Only after [searching for] a long while they found the *oom's* cell phone in the bedroom and at 21:45 they raised the alarm by phoning the neighbours who called for help on the disaster management/BBV (*burgerlike beskermingsvereniging* civil protection society) radio. The area's farm patrols were deployed. The SAPS reacted swiftly. ER24 (an ambulance service) quickly made it to the scene. The *oom* was treated for a cut on his hand. It was later observed that one of the dogs was poisoned.
>
> People we live to comfortably and do not realise what is happening. The *oom* made a mistake, by leaving his house to inspect. Thank you to the *tannie*[3] who kept her wits about her. Thanks to her both lives were saved. The previous night about 15 kilometres away there was also a farm attack.
>
> (Author's translation, JB Marks Security Facebook group,
> 6 September 2017)

The following point has to be made as delicately and precisely as possible to avoid any misconceptions that may be experienced as hurtful or insensitive by potential readers. Instances such as the one cited earlier are no doubt extremely traumatic. They often result in a loss of life and in some cases this does involve unrestrained violence or torture. Life has inherent value. Similarly, freedom from pain and suffering, whether from the deprivation of basic needs or from physical pain inflicted on the human body are aspects inscribed in Human Security and CSS frameworks (cf. Alkire, 2003; Booth, 1991:319). One cannot deny people the right to anger, much less fear in light of cases such as these. I certainly hope that I will never experience something like this. And if I one day move out of my relatively safe flat on the top floor of a four-storey building into a house, then there is an increased chance that something similar (though of course in a regular house and not on a farm) might happen to me. Farm attacks and other forms of crime, such as armed robbery, including house robberies, are a major problem in South Africa. The point I want to make, however, is that we have to disentangle the threat of crime and violence of some forms of crime from notions of persecution. If we do not do this, then we are endorsing the escalation of tensions in an already extremely tense polity. The long quotation shared in the previous paragraph suggests that material gain was the motive for the attack. The robbers left with cell phones. These and other portable electronic devises along with jewellery and other small valuable items are the typical targets of housebreakings and robberies in South Africa. Another anecdote might be worth noting to elaborate the point that much property related crime, which sometimes turns to crimes against the person when they happen to be in the way, is motivated by a quest for these highly liquid items. A friend of mine returned to her flat in 2019 after being away for a few days. She found that all her electronic devices, such as her flat screen television and sound system were gone. She appeared to have been the victim of an organised group that targeted various homes in her neighbourhood in a short space of time. Curiously, her guitar retailing for over ZAR 30,0000 and her guitar amplifier, also worth thousands, were not touched. Much crime is motivated by the acquisition of very specific highly liquid items. Everyone needs a cell phone and everyone wants a television. There are also elicit markets for housebreaking and robbery equipment such as guns. However, not everyone needs a guitar. It would appear that the majority of farm attacks and house robberies do not deviate from this pattern.

Notions of persecution are further fuelled by doomsday-like prophecies shared on social media.

> The tipping point in crime! (JB Marks Security Facebook group, 13 June 2017).

and

> This country is making me mad. The innocent constantly have to sharpen up (their defences), fork out money, make plans . . . but this rubbish are allowed to walk around, commit crimes . . . even murder! AND ACTUALLY NOTHING HAPPENS TO THEM!!!! BECAUSE THEY MAY. Now I am probably going to be removed [from this group], because here we have to be politically correct. Show this to those big men, because this is how we feel.
> (Author's translation. Capitals in the original Afrikaans, JB Marks Security Facebook group 29 June 2017)

The authors of these posts clearly feel quite distressed. They are exhausted by crime. Once again, I do not want to judge this frustration. Still, if posts such as this are read together, then there is an emerging sense of victimhood crystallised on account of a mystified often undefined inhumane other. Although individual quotations might not always do justice to the larger whole, these are merely examples of broader trends in the JB Marks Security Facebook page which potentially fuel notions of persecution and as such possibly of divisions between citizens. Again, there is a threat to a supposed 'in-group' that is ubiquitous and creeping. By its nature the presence of an 'in-group' implies an 'out-group', both of which are the result of logics of equivalence. This out group or criminal identity is mystical precisely because it is based on an unacknowledged logic of equivalence. The inverse, the threatened identity, is somewhat more explicitly defined, though still lacking because it is based on assumed associations. Commentators who link crime to the criminal individual and an implied group to which he belongs only, and not to the social order that reproduces criminality, exacerbate divisions. In the process, whole populations; potential allies in the fight against crime, and partners towards a better South Africa, are discarded.

In concluding this section, I should reiterate the evidence of an enduring gap or distrust between society and the SAPS. As mentioned in Chapter 2, residents often encourage each other through the JB Marks Security Facebook group, to contact a PSC and not the SAPS, when they see a crime has, or is possibly about to be committed. Such distrust might be on account of the notion that the status quo favours an 'other' as opposed to a purported righteous 'protagonist'. Alternatively, it might simply be because there is little trust in an under resourced SAPS. Reliance on PSCs, and as such affirmation of the gap between the SAPS and society, also occurs when group administrators share information from PSCs. More recently this has been negated somewhat by the content, including safety tips, shared by the new station commander.

Apathy and discontent

Community policing forums in Potchefstroom, be it the structure in its entirety or sub-forums have been subject to periodical decay. If the benchmark of a functioning CPF is the presence of a representative from each neighbourhood engaging with the larger meetings on matters of crime, then I have never seen the CPF fully functional. One CPF member of 11 years and an office bearer for much of that time, told me that the station commander contacted him in 2008 to help resurrect the entire structure, which had collapsed completely (Interview, CPF officer bearer, February 2020). Despite advertisements and other efforts to recruit people, it turned out that only those the station approached directly to assist were elected, because there were no other options. I too have witnessed a sub-sector forum completely collapse to the extent where I found myself in a meeting that included only the chair, the secretary and myself. By the stated measure, that each neighbourhood in the sector should have at least one representative in the sub-forum, that meeting was *at least* seven representatives short. Another sector had exactly the same problem a year or two before. In addition to neighbourhood representatives often not showing up to CPF meetings, it is hardly ever the case that each suburb or designated area in need of such representation, has an elected representative in each sector meeting. Consider the following quotation as a case in point.

> The person who is supposed to be the representative seems not to be interested, and the forum will look if they can find someone else to represent the (name of institution) [sic].

(CPF, 2 May 2019)

Both these sectors appear to be up and running again. But, it is clear that there is a significant amount of apathy amongst the public towards the CPF. This represents a great irony. The overwhelming majority of those who complain about crime on social media platforms, especially Facebook, are unwilling to join meetings or neighbourhood watches. This may be understandable in many cases. People work long hours. Meetings are after hours and people want to spend time with their families. Moreover, some might consider their presence at home to be more important to the security of their families (Interview with Oewersig resident, 15 November 2018) than attending a CPF meeting at night or joining a neighbour-hood watch. To expect working people, which the overwhelming majority of residents in this part of the city are, to drive around at all hours of the night, might be asking too much. It is therefore quite logical that CCTV cameras monitored by PSCs are becoming increasingly appealing. They appear to be more efficient. There are neighbourhood watches in Sector One and Sector Two. The neighbour-hood of Van der Hoff Park is one case in point. However, a neighbourhood watch cannot be active 24 hours per day. Compare the following entry in the minutes of a Sector One meeting.

> Although the Neighbourhood Watch Patrol vehicles was trying their best by driving from 22:00–02:00, the burglars also changed the modus operandi to break in before 22:00 or after 02:00 [sic].
>
> (CPF, 2 May 2019)

As mentioned, there is an increase in CFP patrollers joining Afriforum and their neighbourhood watches. Therefore, although there clearly is apathy, there is also discontent with the limits of the CPF and seemingly its association with the SAPS and constitutional rights. This is dangerous, as Afriforum, have a (be it contested) reputation of protecting white minority interests. And as previously mentioned, there have been allegations of their neighbourhood watch occasionally overstep-ping their mandate.

There has been a lot of discussion in CPF meetings about getting other municipal departments in CPF meetings to help deal with matters under their jurisdiction that have a bearing on crime prevention. However, from attend-ing a number of CPF meetings and having access to the minutes of numer-ous other meetings, there is no evidence of this taking place to a meaningful degree between 2018 and 2019. Moreover, the suggestion that service delivery issues not directly related to security are raised at some CPF meetings has also been borne out in the case of Potchefstroom. For example, potholes have at times been mentioned under the standing item on the agenda of 'crime preven-tion through environmental design'. Articulating service delivery issues such as these onto the problem of crime arguably adds to prevalent notions of 'us' and 'them', which as with neighbourhood watch members joining Afriforum's patrols, might pertain to disillusionment with a perceived inept and even hostile state. In many cases, this divisiveness occurs side by side with repeated negli-gence by the public in protecting their property. Despite the efforts of the CPF and its many collaborations with the SAPS and PSCs to sensitise the public to contemporary threats and to offer safety tips, the public at large often remain apathetic to the possibility of falling victim to crime. Despite the CBD being considered a prime spot for theft out of motor vehicles many members of the public still do not lock their doors when they are driving in and around this part of the city (CPF, 2 May 2019).

On the positive side, there are attempts towards evidence-based community policing, a key trait of COP. One example is serious of collective patrols including members of the public, the SAPS and other organisations of particular neighbourhoods prone to housebreaking and theft, to identify hotspots where perpetrators hide immediately after a crime before dispersing. It is also encouraging that there is indeed renewed energy from a new CPF management team. However, as I have mentioned earlier, the functioning of the CPF and its sector forums tend to ebb and flow. Their functioning is highly dependent on the energy levels of key individuals, which may wane over time. This also holds true for the executive committee of the CPF for the entire precinct. Consider the following quotation from a sector meeting on 6 June 2019 (CPF, 6 June 2019):

> Mr (name excluded) indicated that the CPF Exco [Executive Committee] was supposed to have a meeting on the third last Thursday of May, but nobody arrived to attend the meeting, no representative of the police was available. Therefore, no feedback is available from this meeting. Mr (name excluded) also indicated that he has tried to get the previous minutes, but no minutes was sent to him to report on this either.

Hence, there is evidence of already declining interest by August 2019. This is further evidenced by the increased brevity of monthly inputs from neighbourhoods, although it must be said that the minutes for these meetings were still more substantial than any minutes taken in 2018 (cf. CPF, 3 October 2019; CPF, 1 August 2019). During that month's Sector One meeting most neighbourhoods did not have members present. Because of its dependence on individuals who have time, energy and stamina to commit to its functioning for a protracted period of time community policing remains a difficult proposition in Potchefstroom. Some have committed over a decade of their lives to this structure. Some selflessly administrate social media platforms, which I can attest is a very time consuming and exhausting undertaking. I too was an administrator of a general Potchefstroom Facebook page unrelated to crime. At times sifting through the mass of content and removing racist or dealing with otherwise offensive and divisive content can be intellectually and emotionally draining. Yet there are individuals who have taken these types of tasks upon themselves. A more democratic social order based on meaningful participation and praxes requires more of these people who embrace such progressive subject positions.

Conclusion: towards more productive discourse

There are longstanding CPF members who have done extraordinary work to improve safety in Potchefstroom, without reverting to or overtly supporting divisive language or practices. Unfortunately, divisive ideas and practices have at times been articulated onto the very honourable projects and platforms initiated under the rubric of crime reduction.

Problematic discourse and signifiers such as the 'persecution of the Afrikaner', the link between race and crime and the SAPS as an adversary should be dealt with in different ways. The first can simply be revealed and called out as a divisive appendage to meaningful engagements with crime. These voices need to be consistently counter-balanced in the public discourse, preferably by residents with whom they share a language and skin colour. The problem of race, as a descriptor of criminals or suspects is more difficult to deal with. Some residents might read into the descriptor 'black' a subtext and indeed it might often suggest a divisive subtext. Race as a descriptor should therefore ideally be used only where it can obviously serve a purpose in crime fighting. It is however imperative that administrators of the JB Marks Security Facebook page find ways to counterbalance such descriptions of suspects, with greater contextualising articulations of the crime problem and its association with race amongst other aspects. In short: criminality does not equal being black. The CPF or SAPS might more frequently share content on the socio-economic and macroeconomic dimensions of crime.

Regarding the relationship between the SAPS and the public, some progress has been made through greater participation by the SAPS on CPF social media platforms, and other activities, such as mass patrols. It would however appear that senior SAPS officials are stretching themselves to the limit in order to sustain these initiatives (Interview with former SAPS member, January 2020). Constructive engagements with an often-hostile public must continue. What the latter point also indicates, much as with public participation in the CPFs, is that a lot of the successful engagement between SAPS and society depends on specific highly motivated individuals. There is always a real prospect that these productive relations, currently at a foundational stage, might collapse before they truly come to fruition.

Finally, one might view a notion of timidity or impotence on the part of the CPF neighbourhood watches as problematic, as members migrate elsewhere. What is needed is that the SAPS continuously articulate and impress upon all neighbourhood watches the difference between appropriate vigilance and stepping outside of the bounds of democratic sensibilities.

Notes

1 Oom directly translated means uncle, but the Afrikaans version of the word typically extends beyond familial ties and to the best of my knowledge it is more commonly used as such by white Afrikaans South Africans, than people in most other contexts. Oom refers to an older man. It is a designation that suggests respect based on age.
2 *Skelm* means thief or criminal.
3 *Tannie* means aunt. As with *oom*, *tannie* often does not imply familial ties.

References

Alkire, S. 2003. A conceptual framework for human Security. Working paper 2. Centre for Research on Inequality, Human Security and Ethnicity (CRISE). Queen Elizabeth House. University of Oxford.

Boettke, P.J., Lemke, J. and Palagashvili, L. 2016. Re-evaluating community policing in a polycentric system. *Journal of Institutional Economics*, 12(2), pp. 305–325.

Booth, K. 1991. Security and emancipation. *Review of International Studies*, 17(4), pp. 313–326.

Clark, W. and Minnaar, A. 2018. Rural crime in South Africa: An exploratory review of 'farm attacks' and stock theft as primary crimes in rural areas. *Acta Criminologica: Southern African Journal of Criminology*, 31(1), pp. 103–135.

Cordner, G.W. 2010. Community policing: Elements and effects, in R.G. Dunham and G.P. Alpert (eds.) *Critical issues in policing*. Long Grove, IL: Waveland Press, pp. 432–449.

Demirkol, I.C. and Nalla, M.K. 2019. How does police culture shape officers support for community policing. Policing *and Society: An International Journal of Research and Policy*, 29(2), pp. 692–705.

Gill, C., Wesiburd, D., Telep, C., Vitter, Z. and Bennet, T. 2014. Community-oriented policing to reduce crime, disorder and fear and increase satisfaction and legitimacy among citizens: A systematic review. *Journal of Experimental Criminology*, 10, pp. 399–428.

Maguire, E., Kuhns, J., Uchida, C. and Cox, S. 1997. Patterns of community policing in nonurban America. *Journal of Research in Crime & Delinquency*, 34, pp. 368–394.

Moore, M. 1992. Problem-solving and community policing. *Crime and Justice*, 15, pp. 99–158.

Morabito, M.S. 2010. Understanding community policing as an innovation: Patterns of adoption. *Crime and Delinquency*, 54(4), pp. 564–587.

Nalla, M.K. and Newman, G.R. (eds.) 2013. *Community policing in indigenous communities*. London: Routledge.

Palmiotto, M.J., Birzer, M.L. and Prabha Unnithan, N. 2000. Training in community policing: A suggested curriculum. *Policing: An International Journal of Police Strategies & Management*, 23(1), pp. 8–21.

Reisig, M.D. and Giacomazzi, A.L. 1998. Citizen perceptions of community policing: Are attitudes toward the police important? *Policing: An International Journal of Police Strategies & Management*, 21(3), pp. 547–561.

Rosenbaum, D. and Lurigio, A. 1994. An inside look at community policing reform: Definitions, organisational changes, and evaluation findings. *Crime & Delinquency*, 40, pp. 299–314.

Sampson, R. and Raudenbush, S. 1999. Systematic social observation of public spaces: A new look at disorder in urban neighbourhoods. *American Journal of Sociology*, 105, pp. 603–651.

South Africa. 2001. *Interim regulations on community police forums and boards*. Pretoria: Government Printers.

South Africa. 1997. *The criminal procedure act 51 of 1977*. Pretoria: Government Printers.

South Africa. 1995. *The South African police service act 68 of 1995*. Pretoria: Government Printers.

Van Riet, G. 2020. Intermediating between conflict and security: Private security companies as infrastructures of security in post-apartheid South Africa. *Politikon: The South African Journal of Political Studies*, 47(1), pp. 81–98.

Van Riet, G. 2016. The limits of constitutionalism and political development. *New Contree*, 75, pp. 98–115.
Vitale, A.S. 2017. *The end of policing*. London: Verso.

Interviews

CPF (Community Policing Forum) office bearer. Interviewed by Gideon van Riet. 4 February 2020. Potchefstroom.
Oewersig resident. Interviewed by Gideon van Riet. 15 November 2018. Potchefstroom.
Police officer. Interviewed by Gideon van Riet. 9 August 2018. Potchefstroom.
Retired police officer. Interviewed by Gideon van Riet. 24 January 2020. Potchefstroom.
Retired police officer. Interviewed by Gideon van Riet. 6 July 2017. Potchefstroom.

Minutes of meetings cited

CPF Sector One. 2019. Minutes of meeting held on 3 October 2019 at 17:30. SAPS Boardroom.
CPF Sector One. 2019. Minutes of meeting held on 1 August 2019 at 17:30. SAPS Boardroom.
CPF Sector One. 2019. Minutes of meeting held on 6 June 2019 at 17:30. SAPS Boardroom.
CPF Sector One. 2019. Minutes of meeting held on 2 May 2019 at 17:30. SAPS Boardroom.

Facebook posts

The full details for each post are supplied in each in-text reference. These references include enough details for interested parties to confirm the citation. However, supplying additional information here, would mean linking comments/quotations to names, which would contravene the conditions set by the group administrators for usage of this data.

5 Trauma, debate and the death drive

Discursive entanglements of crime reporting in the Potchefstroom Herald

I was in grade two, when PW Botha suffered a mild stroke and FW de Klerk took over as the preliminary and then permanent leader of the NP and as the president of South Africa. I remember this very well. Because of rather protective parents, I developed interests that did not require me leaving the house. Consequently, I was quite aware, for an eight-year-old, and then continuing throughout my childhood, of the politics, social, and economic dynamics within South African society. By this of course I mean, at least for someone from my positionality. I deliberately use the term positionality here, as Abboud et al. (2018:280) argue that critical approaches to security require consistent reflection upon one's positionality. Just because I disagree with many with whom I share an 'ethnicity', that does not render my identity completely discreet from this loosely sutured group.

In the early 1990s, it would have been abundantly clear to any outsider that white South Africa as whole had to make a mind-shift, a major one. Realistically, such a mind-shift would never be linear. However, 30 years later, many would be quite disappointed. In the early 1990s the rules by which society from the global level to our particular local level were constituted and reiterated were changing at a rate of knots. Familicide and a satanic panic – likely influenced by similar discourses in North-America as source of cultural information – were some of the ways white South Africa 'dealt' with the angst of their changing reality. Policemen, it appeared, were committing suicide at an unprecedented rate. The reasons possibly related to the many of the deeds they were expected to commit. The talk amongst older white adults, throughout my primary school years, which coincidentally ended in 1994, was that 'the country' (read their/our country) was being 'given away' and at times that there was a looming 'blood bath' between racial groups.

We know that notions of de Klerk as a peace-maker should be balanced with the limited to no alternatives he had and then there is the persistent stigma of the alleged 'third force' armed by the state to clash with ANC supporters and supposedly designed to weaken the ANCs bargaining position during negotiations (see Van Riet, 2016). Throughout my primary school years, I watched the news most nights. I saw the tug of war during negotiations between the ruling NP and the ANC. I saw footage of the Boipatong Massacre, which was such an attack

DOI: 10.4324/9781003028185-8

on ANC members. I saw the rise in support for the neo-fascist AWB, and their storming into the negotiations venue in Kempton Park, East of Johannesburg, with an armed vehicle. I also read magazine articles on the 'battle of Ventersdorp' where the Police clashed with this group and television footage of the 'battle of Bophuthatswana', where the AWB entered this particular homeland area to supposedly protect its 'sovereignty' – an apartheid creation. I recall the moment on that day when a Bophuthatswana soldier executed three AWB 'soldiers' live on global television. And then finally, there was surely one of the most indiscriminate, ill-conceived and irresponsible act in the history of South Africa; the murder of Chris Hani, the leader of the South African Communist Party (SACP), shortly before the planned elections. The murder was orchestrated by a member of the now defunct Conservative Party (CP). On that particular night de Klerk's impotence in affecting the future of South Africa any further was quite clear. He had to put Nelson Mandela on the air. Hopefully Mandela could calm the people down. This he did, at least sufficiently.

In living memory there has always been tension in society, the severity of which has ebbed and flowed. In the early 1990s, this tension was nagging, palpable and traumatic. It never went away. It could not, for reasons pertaining to relative material depravation and incessant psychological harm throughout the population. In the process the problem of crime has arguably become intertwined with this tension, which flared up again under president Zuma's ill-fated reign.

*****

Introduction

Following the aforementioned personal account this chapter approaches the idea of an echo-chamber linked to crime discourse through the optic of trauma and based the media analysis. Although the *Potchefstroom Herald* is not a national newspaper, let alone a national television or radio channel, being confronted by trauma, including those of others, incessantly in smaller or larger doses, gnaws away at you, regardless of your positionality in society. Of course, the tension experienced by a lower-middleclass white boy and later by a properly middle-class white adult man, is far removed from the struggles experienced by most other citizens. I would argue that complex trauma, in the psychological sense (see Luxenburg et al., 2001a, 2001b), structural violence (see Farmer, 2001; Galtung, 1969) and 'wretched' existence for many (Fanon, 2004 edition), remain appropriate tools that shed light on much of South African society (see Van Riet, 2017, chapter 3). My research in Ikageng is continuing. I hope to publish meaningful qualitative data-based insight on the experiences of residents served by that police precinct as a matter of extreme importance. This book, however, which started out as a research project with no budget, deals primarily with the responses of the largely white middle and working classes in Potchefstroom to the stimulus of crime. Based on initial insights from fieldwork in Ikageng I am, however, better able to link these two contexts and the important relationship between them. Following a non-unitary Laclauian understanding of subjectivities, the chapter continues the argument that groups based on labels such as race, class and language, as identities often portrayed in South Africa, are not homogenous and that differences within groups are numerous. These differences, even in a relatively conservative context such as Potchefstroom are sources of opportunity for progressive change. As such, part of the potential praxes in the service of radical and plural democracy espoused by Laclau and Mouffe (2014 edition, see chapter 4) might include getting people historically separated, but who share similar concerns, to talk to each other.

Reflecting upon my childhood experiences, have made me somewhat less judgemental of the responses many people have to unremitting reports of crime. Incessant fear combined with other forms of tension and uncertainty is taxing. But, this does not mean that we do not still have to disentangle divisive ancillary discourses from the realities of crime and the fundamental underlying political economic causes of crime, as conceptualised in Chapter 1. Such causes directly and indirectly implicate these very (fearful) elites. The situation is complicated and, as such, distinguishing the real problem of crime from unwanted ancillary discourses is even more important.[1] Even though I still disagree with many of these responses to the stimulus of crime reporting and crime discourse in general, empathy or even just sympathy for people you disagree with is an important principle of research ethics. Furthermore, such sympathy is quite necessary in a context where tensions run high and have actually flared up in recent years. The point should be to get different people, strategically using logics of equivalence and difference and eventually strategies of construction to work towards a shared vernacular and towards approximate shared objectives (Laclau and Mouffe, 2014[1985]:172–174). Interventions towards such ends might be quite difficult in the absence of mutual trust, stemming from a recognition of each other's fears and doubts.

This chapter analyses the *Potchefstroom Herald* as a security infrastructure from the period 1 January 2018 to 31 December 2019. As the reader is aware, the chapter sits between two other chapters on similar infrastructures which direct the flow of crime discourse, the CPF, primarily through its monthly meetings and Facebook page, dealt with in the previous chapter, and the Facebook pages of PSCs, which will be dealt with in the following chapter. All three these infrastructures are to some extent related. They cross-post and draw otherwise on each other's content. This makes the *Herald* far more impactful than the 4,895 copies it sells per week, as measured for the last quarter of 2019, would suggest. The fact that these sales are a decrease of 3% from the previous quarter (Breitenbach, 2020), also reflects an important piece of the problem this chapter addresses: the economics of newspaper publishing. We will get to that in due course. The *Herald* can also be read online and selected articles are shared on the JB Marks Security Facebook group and on a general Potchefstroom Facebook group. The public depends on the latter platform for all sorts of communication, such as asking for advice on a range of matters or posting adverts.

This chapter takes the *Potchefstroom Herald* as its unit of analysis. As for usual this chapter too draws on post-structural hegemony to explain how hegemony is reproduced and how it can potentially be rearticulated. However, in addition to the twin theoretical pillars of infrastructure and hegemony, the chapter also draws on Lacanian psychoanalysis. This form of analysis is significant to the central argument of this chapter in the following way. Lacan talks of the symbolic orders of signs and signification into which children are socialised. These symbolic orders vary between contexts. *The central argument of this chapter, based on a reading of the Potchefstroom Herald, is that not all white Afrikaans speaking residents of Potchefstroom are located in a (loosely sutured) symbolic order amenable to constructive dialogue towards radical and plural democracy. However, others who are potentially better placed, should be more involved in constructive praxes towards these ends.*

The rest of the chapter is structured as follows. After some theoretical notes on Lacanian psychoanalysis and its purpose in this chapter in the following section, the section thereafter outlines in brief the structure of the typical weekly edition of the *Herald*. My objective with this section is to orientate the reader on how the newspaper is typically laid out and the amount of content it is able to produce on a weekly basis. Thereafter, the chapter will deal with the debates and topics that constitute the typical content of the newspaper. This discussion is important in explaining how the newspaper at times reproduces a given order of things, simply conceived of as a social structure that residents have become accustomed to. Reproducing or rearticulating in Laclauian terms, such a social order, may fuel notions of 'us' and 'them' which are often inflected with the language of 'social decay'. Thereafter, the chapter revisits the topics of trauma and psychoanalysis and how these are seemingly borne out in some of the debates we see in the newspaper. With the problematic of crime reporting and social divisions outlined, the chapter concludes by reporting on and considering potential interventions.

Lacan

The choice of Lacanian psychoanalysis in this chapter as opposed to Freud, for example, is on account of clear compatibility with Laclau's work. Both authors emphasise the notion of 'a lack' in language or its inability to fully represent reality or a perceived reality. Indeed, Laclau was influenced by Lacan in this regard, in the sense that the supposedly innate relationship between signifiers and the signified, are actually contestable. Both agree, there is no incontestable signified. As other authors have, however, mentioned (see Cho, 2006:19; Homer, 2005:3;12;36;46;51; Feldman, 1987:102), it would appear to me that there are many commonalities between Lacan and Freud. Indeed, initially in his career, Lacan focussed on revising Freudian psychoanalysis. Having encountered the work of Lacan, however, I have found his ideas befitting of the argument I wish to make in this chapter. Lacan is used in broad strokes and deliberately simplified, to help explain the mind shift from an extremely conservative perspective or world view to one that is possibly more open and amenable to productive interaction across longstanding divides.

The Lacanian shift at early childhood from the illusionary pre-linguistic state to the symbolic level where the unconscious is structured like language, occurs in a context of a traumatic Real. Lacan argues that between 6–18 months a child encounters herself in a state of mirroring. She realises that she is different from the world and her mother, the first object of the child's narcissistic attachment (Feldman, 1987:104). It is at this stage where an unconscious is formed, which is structured *like* language. As such, it is 'a formal system with its own rules and regulations which could not be infringed but at the same time remained unconscious to the individual system users' (Homer, 2005:35). This subconscious language-like system of thought is also referred to as the symbolic order. The symbol of the father is the original 'no' uttered to the child and the child finds throughout his/her life new objects of desire to fill a void, but, it is a void that cannot be filled (Feldman, 1987:104). This reading of the Oedipus complex is quite different from Freud. For Lacan, the Oedipus complex is a signifier, not a signified (Feldman, 1987:103). The father and the phallus are mere symbols, animated by the aforementioned original 'no' (Lacan, 1961 seminar 22:10). The person encounters the Other (capital 'O'), the symbolic order, of which she is merely part, but it is not quite clear how.

To alleviate the resulting angst another other (lower case 'o') is created. This lower case 'o' is what Lacan calls the *objet petit a*. These are objects of desire whereby the person tries to position herself in the relation to the Other. It is one way of answering 'who am I?' But, the other/*objet petit a* changes frequently and the essential drive towards potency (phallus) cannot ever be fulfilled, as the phallus was a fake, a mere symbol in the first place. The real father too was incomplete in the same way. As such, the primal 'lack of' is the *idea* of the father or phallus/potency, which was never real, but, it is, the basis for the fundamental drive throughout life. The primary drive, what Lacan calls the death drive, is revealed when there is an evident break in the symbolic order inhabited by the individual, where the gap between signifier and signified is clearest. Here fantasy as a realm, which never disappeared since early childhood, reveals itself. Fantasy, however, does not entail the imaginary ever succeeding in obtaining an object of desire. Its satisfaction lies in the setting it invokes where *jouissance*, or the simultaneous revelling in pleasure and pain associated with the object of desire manifests (Vincent, 2020:54–55; Laurita, 2010:67; Lacan, 1961 Seminar 151:3).

Neither the pre-linguistic fantastic order, which is not a stage but a continuing order, nor the symbolic order of incomplete semantics (similar to Laclau) can contend with the Real. The Real, is the elusive reality beyond language and perception. Technically the Real is not real, based on a Lacanian reading, as it falls outside of the symbolic order of signification, which he equates to reality (Lacan, 1960 Seminar 1:37–38;46). Nevertheless, it is experienced through trauma that invokes the realm of fantasy and often the death drive of self-defeating repetition in a quest for obtaining the unobtainable (Homer, 2005:83–84).

I have thus far identified an induction into the symbolic order largely derived by Lacan from structural linguistics at a young age. This order differs depending on context, but offers the inductee some cues as to the nature of the order and as such

the Other. I would like to argue that there is, to simplify, for the sake of explanation, an additional mind shift needed for many in contexts such as South Africa.
Lacan did not to the best of my knowledge deal with such a double move. Once a
very young child is inducted into a symbolic order and learns symbolic usage, she
becomes socialised into a context, where some symbols, words and practices are
more acceptable than others. It is learned from a young age that drawing on and
working with some symbols will achieve better results than drawing on others.
The problem in a context such as South Africa is that many of those who lived
during apartheid or even its immediate aftermath needed do undergo various itera-
tions of resocialisation. For simplicity's sake, we might call this a 'double movement', first into a symbolic order of socialisation through the family, which for
many, especially older people, has largely been socialisation through conservative
institutions, such as schools. Another second shift into a more open and inclusive
and potentially more productive symbolic order is required.

It is at this level, of the particular nature of symbolic orders, where I find differences in Potchefstroom. Not everyone is alike, even if they share a similar
position within common matrices of categorisation such as race, class and language. The nature of the symbolic order inhabited by the individual, has a bearing
on how she responds to the stimulus of crime and other reporting in the media.
It is at this level that constructive intervention might be very difficult, but also
very necessary and security infrastructures such as the *Herald* can have a positive
influence. Unfortunately, the limited funding possible for such a small newspaper,
as we will see, limits its ability to promote a more progressive symbolic order.

The Herald: layout

Up to 2019 local shops and street vendors sold the *Herald* every Thursday. It has
been distributed for free since 2020. The paper is never particularly thick. The
newspaper proper, is typically around 20–25 pages in length. There is usually
a substantial number of pamphlets and other promotional material from local
retailers placed in the middle. These kept the price of the publication to just
under ZAR 6 in 2018 and just over ZAR 6 in 2019. A very limited number
of staff members produce the *Herald* and its sister publication the *North West
Gazette*, which is distributed on Tuesdays through mailboxes free of charge. The
latter publication is even more focussed on advertisements and as such I did not
consider it particularly significant to this research. The *Herald* typically only features contributions from about four journalists, one of whom almost exclusively
focusses on sport.

The typical layout of the newspaper includes a title page with a set of head-
lines and titles of important stories found inside on particular pages. There is
also a single main headline on the front page along with the name of the newspaper and some advertisements. In a number of editions of 2019 the very top
of the cover page included a running head sponsored by an opposition party. It
reads *Slaan terug, daar is hoop*. This may be translated as 'hit/fight back, there is
hope'. The next number of pages, up to about page six or seven would typically be

filled with stories on crime and municipal service delivery. Included is a weekly column by the executive mayor of the city. There might also be articles dealing with complaints by the public on services and rebuttals by municipal representatives. On page seven or eight there is a weekly section called *Hoe sê Ore* (What does Ears say?). 'Ears' is an imaginary character created by the editorial staff that engages in a type of light hearted gossip. Ore often complains about the same types of issues found in the preceding pages. Typically, on the same page there would be a weekly column for members of the public to contribute, entitled *Kommentaar* (Comment). The column would normally (though not always) have a sub-heading, specific to the commentary being given. At the bottom of the same page a weekly column is filled by a message from a Christian minister, preaching from the Bible and offering advice in 'difficult times'. Following this page there is typically a section of printed text messages sent by members of the public to the newspaper. This section covers about a third of a page. These text messages are often quite negative about the state of crime and municipal service delivery, although promisingly there are occasionally messages of praise, including praise for public servants and even the SAPS for noteworthy service delivered in particular instances. Thereafter, other human-interest pieces, such as articles on the passing of notable members of the community and local events takes us to a list of job adverts (around page 14) and the classifieds section. These sections would fill the paper up to about page 20, from where local and regional sport fills the final pages.

The Herald: content

As mentioned earlier, most of the main news content in the *Herald* for the study period can be reduced to two broad topics. They are crime and municipal service delivery. A simple frequency analysis of articles using Statistical Package for Social Scientists (SPSS) for the first six months of 2018 revealed that these two topics were the subject of over 61% of articles, excluding the sport pages. There are likely reasons for this. The newspaper's limited staff probably does not help. During an interview one journalist told me that he would go to the police station every Monday morning to get a list of incidents (Journalist interview, June 2017). The newspaper then seemingly relies on many of these incidents to fill its pages. Of course, one may argue that crime is indeed a serious problem in Potchefstroom and therefore should be covered extensively in the newspaper. There is some merit to such an argument. My point is, however, that reading the *Herald*, often leaves one with the impression that there is very little else happening in the city. This cannot be true.

Then there is also the spectacle tantamount to an open sparing match between the mayor and members of the public. This near weekly confrontation might be read as a clash between prominent members from the two different ideal-type – political and economic – elites described in previous chapters. It plays out in the newspaper, through the weekly mayoral column and responses thereto largely by the white middleclass members of the public. Alternatively, a political organisation or a citizen might complain about municipal service delivery issues or newly planned housing developments. These complaints would be followed by rebuttals by a municipal spokesperson.

In both these cases the newspaper relies on a type of spectacle to fill its pages. We have already deduced a lack of staff capacity from the previous section. A disproportionate focus on crime and municipal services in the newspaper, is often perpetuated by the numerous reserved columns and spaces for external content mentioned earlier. The *Herald* reports weekly on housebreakings, house robberies, theft of and out of motor vehicles and occasional instances of common robbery in Potchefstroom. These are augmented by frequent reports of more violent crimes, such as murder and rape in Ikageng. This does not mean that theft and housebreaking do not occur in Ikageng. Indeed, crime statistics suggest otherwise (see CrimestatsSA, 2019). It does however seem that someone reading the *Herald* frequently, would get the impression of Potchefstroom as the locus of property crime and that Ikageng as the locus of violent crime, when the situation is more complicated. This once more opens the door for racist discourse that insinuates notions of violent blacks.

At the same time there are the issues, many of which have merit, about municipal services not being adequately delivered. The reader will recall the financial troubles of the municipality as one contributing factor. Together these discourses of rampant crime and service delivery failure, seem to drive a type of 'failed state' narrative, which implicates the ANC government and at times the SAPS. This is evident in various comments. Consider the following:

> could not care attitude of the mayor, [the] speaker as well as incompetent members of the mayoral committee demonstrates no will to manage the municipality.
>
> (Anon, 2019:6, 9 June; Author's translation)

Then there are numerous complaints over specific services. These include street lights and waste removal not functioning properly and arguments that the municipality is in a state of *agteruitgang* or 'decay' (Anon, 2019:6, 9 June; Anon, 2018a:4, 3 May).

To the failed state narrative, we should add less frequent, but still relatively common, articles on changes on the North-West University's Potchefstroom Campus, where Afrikaans and by association sutured notions of 'Afrikaner culture' are under threat. Under apartheid, the Potchefstroom University for Christian Higher Education (PU for CHE) or PUK after the initial name *Potchefstroomse Universiteitskollege* (Potchefstroom University College), was a significant institution of the Afrikaner establishment. Its alumni include former president FW de Klerk. Therefore, the 'language debate' dealing with the language of instruction as opposed to making universities more broadly accessible to other language groups (see Oosthuizen, 2019; Herman, 2016; Scholtz and Scholtz, 2008) is about so much more. This debate has ebbed and flowed over the past 20 years on various campuses in South Africa, including Stellenbosch University (SU), the

University of Pretoria (UP) and the University of the Free State (UFS). All of these universities now *de facto* teach in English, and there appears to be a concern that the NWU's Potchefstroom Campus is going the same way. When read along with debates about new working class (read black) settlements planned in Potchefstroom, it does appear that language has sometimes been used conveniently, as an excuse to exclude people based on race and the defence of a particular 'cultural group' (cf. Adams, 2018:3). Also consider the following thinly veiled racist text message, 'The PUK is up to date. Purple is the new black' (Anonymous text message, 22 February, 2018:8). This text message follows the replacement of the old PU for CHE's maroon colour scheme with a different shade of purple for the entire NWU, stretching across all three of its campuses. Its author is clearly concerned with racial transformation on the Potchefstroom Campus. In an op-ed Connie Mulder (2018:8) of the Solidarity (widely regarded as a conservative trade union) Research Institute following a change to the university's language policy, argued that a decision to anglicise has already been taken. They (Solidarity) believe the availability of Afrikaans is gradually being eroded. They state outright that they are concerned not just because it is language that will change. For this group English is merely a prerequisite for transformation and 'that is only the beginning'. He goes on to state that 'What will follow are demands for decolonisation, anarchism and the destruction of student life and the radical racial transformation of staff and students' (Author's translation). Clearly commentators such as these and those who are swayed by them, inhabit a specific symbolic order that is completely at odds with radical and plural democracy and most notions of inclusivity. The *Herald* was willing to publish this material equating transformation and decolonisation to 'anarchism' (where the word is clearly not used in reference to the political philosophy) and 'the destruction of student life'. As will be argued in the following section, this very conservative symbolic order is constraining and well outside of the bounds of spaces for productive dialogue.

Crime, as argued throughout this book has been rendered a discursive space amenable to conservative agendas and it has been conflated with conservative ancillary discourses. To reiterate, none of this implies that fear of crime in contemporary South Africa is irrational, nor do I suggest that there is something wrong at a base level with Afrikaans language rights. The point, as with most of this book, is that these articulations become tools for problematic discursive practices and for fear mongering in the service of a totality sutured through intertwined discourses and equivalences. A supposedly homogenous Afrikaner group is under threat from progressive forces (as a totality) that seek 'anarchism' (read chaos) and decolonisation (read the end of Afrikaner supremacy in certain cultural institutions). So we now have the real problem of crime, reported on consistently, where the SAPS, for its limited capacity, are often viewed as an enemy (see Chapter 2). We also have weekly exchanges between representatives of two elite groups, largely defined in this context based on the historical fault lines of race, on the matter of service delivery. To this we may add the at times hijacked discourse on the university's language policy.

Quite disappointingly, frustrated protesters in Ikageng, fed up with far worse living conditions and services when compared to the aforementioned commentators, are often portrayed in the *Herald* and on social media as unruly, 'barbaric' and as a threat (see Smalman, 2019:6). They remain part of the constitutive and neglected outside of a sutured hegemonic totality. This might be in part, because other threats such as crime have been associated with skin colour (Van Riet, 2020), even when such a link, where it might exist is far more complex. The link between race and crime is complex because the same processes that produce relative affluence for some, also feed the inequality and poverty that often, though admittedly not in every instance, drives crime.

There appears to be a massive disturbance in the order of things for many white working class and middleclass residents of Potchefstroom. It would appear that fear begets fear and that each of the previous discourses adds an additional layer to the fear experienced by many. Anxiety over the fundamental disturbance in the order of things, to re-join the discussion from previous chapters, seems to manifest as a new type of perceived onslaught on Afrikaner people.

This contemporary 'onslaught' is one associated with the realities of losing a lot of political power, although definitely far from all of it. The disturbance in the order of things that has gradually been taking place for at least 30 years, has only recently gained momentum in the form of transformation at the university, a major employer in the city and proposed changes in the geographies of race and class through articulated official representations of space. In a sense, some very old, even pre-apartheid 'chickens' are finally 'coming home to roost'. The potential implications for property values, increases in crime and other concerns linked to change, which some are worried about, are the result of centuries of cumulative inequality. It must, however, be emphasised that the many members of the political elite are also quite significantly implicated in the perpetuation and reiteration of this inequality and the widespread multidimensional poverty reported on in Chapter 1. Many within the political elite still rely on the average impoverished citizen to stay in power and the changes some members of the economic elite are so concerned with, are possibly considered necessary for this elite to maintain its position.

Trauma, debate and the death drive

Reading the reporting on crime in the *Herald* can be quite unsettling. While working through two years of newspaper articles for the purposes of writing this chapter, there were instances where I needed to take a break, sometimes even for a few days. If it is this unsettling for me, who ought to have some perspective on the social, economic, political and psychological drivers of this violence, I can only imagine how it might be experienced by those less initiated. Being trained to understand the content of the newspaper ought to help blunt the realities reported on. But no one is immune to constant reports of violent house robberies, rape and murder. As mentioned previously, structural violence and complex trauma over generations, though rearticulated in different ways seem like appropriate perspectives on grasping key aspects of the contemporary South African social order. Perhaps for some of these reasons, notions of onslaught find currency amongst those who have been socialised in a particularly conservative context. Pre-existing narratives can, as it were, be projected onto the stimulus of news articles. It is here where Lacanian psychoanalysis might be useful.

The white and Afrikaans group in Potchefstroom are not homogenous. For example, they do not all support the same political party. Municipal ward councillors come from diverse parties, although for the aforementioned group two apparently stand out, judging on the composition of ward councillors in Potchefstroom, excluding Ikageng, Mohadin and Promosa (JB Marks Municipality, 2018:203–218). Both these parties are arguably at least somewhat right of centre. If we continue to apply a logic of difference, then the sparing fights noted earlier, between the city council and members of the public appears to also reflect the diversity within the white Afrikaans population and as such the symbolic orders inhabited by this diverse group. The most conservative members of society and their representatives are more prone to knee-jerk reactions and full article length complaints over matters of service delivery and desegregation. The frequency of such complaints makes their reactions quite predictable and as such more easily dismissible, as those of unreasonable outsiders. But, they are not outsiders in terms of a constitutive outside of the sutured hegemonic inside, as Laclau and Mouffe would have it. They are very much on the inside and even move in policy circles. They are, however, outside of the bounds of constructive engagement. As such, they are both easily dismissible in democratic deliberations, but also somewhat influential through their relatively substantial following. Luckily, part of this influence can also be explained by an overlap between symbolic orders, which can never be completely bounded units.

It would appear that because this most conservative group often falls outside the bounds of constructive deliberation, that they are involved in the self-defeating drive that is part of the conceptual toolkit of psychoanalysis. In other words, they are engaged in self-undermining activities that are always very likely to fail and which, in the context of local politics, weaken their position. Members of this group dismiss efforts at demographic transformation and desegregation out of hand. They can therefore be easily dismissed as overly conservative and out of touch with the prevailing symbolic order, be it a dynamic and temporarily sutured order. These out of hand, almost reflex, reactions could be interpreted as fantasy, where the satisfaction of the death drive is in the moment of utterance and not in the possibility for any meaningful success. This group possibly finds the status quo most traumatic. There is an angst associated with a loss of relative position reflected in many of the articles published in the *Herald*. The example of the rezoning of parks in various suburbs of Potchefstroom to accommodate housing for working class earners is a good example (see Leshage, 2018:4). These homes will be aimed at people like nurses, teachers, taxi drivers and members of the SAPS and the South African National Defence Force (SANDF). The project was also motivated by the mayor as a way for the municipality to make money, and because parks apparently are 'a haven for criminals'. We should resist engaging with the latter comment for now. What is most significant for the purposes of this chapter is that all political parties, including independent councillors supported this rezoning. The only exception was the party most would consider as South Africa's most conservative. This party's representative stated that the project will overload the infrastructure of 'well planned' (apartheid era) suburbs. It will impact house prices and that 'rezoning of parks is politically driven'. Here it is very difficult not to understand 'politically driven' as code for desegregation.

It may be theorised that members of this group have not made the double movement from the world of fantasy through a conservative symbolic order to a more productive symbolic order. Consequently, their fleeting grasp on the Real is even weaker than that of many others. The Real is traumatic because it is violent and necessarily incomprehensible. It is traumatic for most people who read about physical violence in the city and one might imagine grand corruption in all spheres of government. The Real might possibly be especially traumatic for this group, as they lack the symbolic acumen to disentangle the violence of the Real from a general notion of onslaught on them as a loosely sutured 'people'. For lack of a better term, they seem to lack a realisation that some things cannot be controlled.

In contrast, other white Afrikaans residents of Potchefstroom and likely the supporters of the less conservative of the two parties mentioned earlier, appear to be more pragmatic in dealing with similar issues. They typically do this by focussing on procedure in how change is facilitated, essentially leveraging the fact South Africa is a constitutional democracy. The latter group's intentions may not always be sincere. Nonetheless, they have acquired a vocabulary within a symbolic order, that is more productive and useful to their own ends, as opposed to the death work performed by many within the previously mentioned group. Having acquired this vocabulary and a better sense of the types of action fall within the realms of possibility and which do not this second group has been far more capable of navigating changes within society. The upshot is that in the process they have had far more successful outcomes in regards to the Lefebvrian dialectic between representations of space and spaces of representations, at times successfully effecting partially or more fully exclusionary social spaces. We will deal with examples of these in chapters eight and nine. On the aforementioned rezoning of parks in Potchefstroom, for middle-income housing development, this party's representative stated that they support development, provided it is done according to correct legal procedures and that an appropriate balance between green open spaces and housing. They will ensure residents of municipal wards are aware of public participation. These rules, give them options for resistance, should they choose to use them, but importantly, they draw on more commonly acceptable symbols. As should be clear to the keen observer, based on the previous chapter, this group too is not homogenous. Some are more genuine in reaching across divisions within society, while others have likely merely acquired the symbolic capital to navigate changes in society and resist where feasible. This lack of homogeneity is reflected in some of the content published in the *Herald*, but also from observations during CPF meetings, where people I would classify within this group, display greater empathy towards others. Such individuals are seemingly key in affected progressive changes in society. For this to happen they need to engage more critically with the more conservative members of their group and of course work towards collaboration across elite and other groups.

With the aid of Lacanian psychoanalysis, I have established once more, the categorical outlines of a sutured hegemonic totality in Potchefstroom, including the subject positions of contemporary elite groups, and the masses portrayed, sometimes unwittingly, as the cause of crime. I have also established key aspects of the politics between these groups. Not only is there a division between the elites in Potchefstroom and the rest of the population and between two ideal-type elites, but there are also differences within the economic elite, revealing its essentially sutured nature. Although the less conservative component of the economic elite inhabits a potentially more productive symbolic order, this is not enough, as that position can still be abused towards relatively conservative ends. The following section will more fully bring the Lacanian psychoanalysis used earlier into conversation with the Laclauian notion of hegemony, to derive possible points of interference towards progressive change. In keeping with the psychoanalytic focus of this chapter, we might understand hegemonic rearticulation as interferences at the level of symbolic orders.

Interventions across a complex spectrum

A key point of interference into the injustices of the status quo would be deliberation around the meaning of trauma as it is experienced by diverse members of society. As an implied signifier and actual physical symptom, trauma affects a diverse range of people. Trauma also affects different people in different ways and different people experience different traumas. A key problem with trauma within the context of this study is how it is often intertwined with another notion, that of onslaught on a category such as the Afrikaner, their culture and even their physical survival. In such instances, trauma may be closely related with a self-defeating death drive, stemming from the problematic symbolic order inhabited by a significant minority. The Real as an extra-linguistic and extra-symbolic set of experiences is traumatic and incomprehensible, but some symbolic orders are better able to navigate traumatic incomprehension than others. Empowering people to better navigate the Real, will likely be associated with instilling a realisation of the limits of their power over it and the catharsis of such a realisation. This might free people to some extent to engage more constructively with the problem of crime, but also with each other more generally. For example, signifiers, such as 'barbaric' attached to even more helpless and voiceless protesters in Ikageng could be approached more constructively, through better interaction. Signifiers such as 'Afrikaner' have clearly been demonstrated as problematic for the diversity of people who might be described as such. It is indeed a sutured category.

Building good faith interaction across historical divides as a more united front for example to hold local government accountable or for social and economic investments, which might reduce the root causes of crime, would require more than merely transitioning between a binary of symbolic orders. Not only, is the binary and the double movement described earlier a deliberate oversimplification for the sake of initially illustrating a point, but it is also true that further, even perpetual, symbolic shifts are required and will have to be accompanied by real action. As such, the instability in the hegemonic order, as far as attitudes in Potchefstroom are concerned, can best be exploited by progressives who have made the 'double movement', hopefully in collaboration across historical divisions, thereby, swaying firstly, others within this broad category (of 'double movers') and then the community at large. Fantasy and death drive along with symbolic orders form part of the ontological component of the laager. Therefore, getting the most conservative elements to transition will likely be the most challenging, as their socialisation is most removed from the realm of productive possibilities. Consistent challenges by those who look like them and speak their language seem the most logical approach towards such change.

The iterative, incremental, progressive change envisaged here should include meaningful, respectful and accommodating interaction with potential allies in Ikageng and within the political elite, which as a category is, by definition, also is not a homogenous group. Moreover, this division within the political elite is evidenced in the tumultuous internal politics in the city council in recent years (see Maphanga, 2020). I will conclude this chapter by highlighting some evidence of potential for progressive change, as evidenced in the *Herald*.

Over the past year an associate professor in Town and Regional Planning at the North-West University has been contributing a column to the newspaper focussed on 'more positive news'. It seems the motivation behind this intervention was the barrage of sad stories that have been filling the pages of the *Herald*. Her columns, have focussed on diverse aspects from a plea for recycling, framed in terms of making a constructive difference, to comments over race relations and an acknowledgement of the far greater problems many residents of Ikageng face, when compared to Potchefstroom (cf. Cilliers, 2019:11). Initiatives such as this, especially when operating at a greater scale, have the potential for interrupting symbolic orders, by using the language of 'we' as opposed to 'us' and 'them'. Ideally articulations such as 'we' may become hegemonic in the longer term. In other instances, residents of Potchefstroom have demonstrated compassion and agreement across historic lines of division. Consider the following comment on solidarity with protesters.

> (I) agree with protesters, the mayor should go. He frequently makes promises he does not keep.
>
> (Anonymous text message, 9 August 2018:8.
> Author's translation)

Other positive examples, as mentioned previously, include columns in the commentary section and published text messages that praise the SAPS or other public servants. These comments, should be encouraged as they help restore balance to the trauma often affixed to the weekly editions of the *Herald*. Consider the following examples:

> I just want to say thank you to the friendly people who assisted us so quickly with our applications for new ID (documents). Us, the two old ladies. Keep well.
>
> (22 February 2018:8. Author's translation)

The new police station commander appointed in 2018, has clearly done much to improve the relationship between the public and the SAPS. As mentioned in the previous chapter, this has included frequent information sharing, but his efforts have also included efforts for collaboration between the SAPS and the public beyond the CPF. He has for example launched more deliberative and problem solving exercises, such as the 'crime summit' of 2018, where the public was invited to a half day workshop where they could raise their grievances and collaborate towards possible solutions (Wetdewich, 2018:4). To this, one resident responded positively, stating that 'It is heartening that residents' inputs are requested for a crime action plan. The level trust between the community and the police is probably at an all-time low' (Anon, 2018b:6).

Finally, constructive engagement, based on genuine deliberation and collaboration and as such radical and plural democracy, will also entail those in hitherto powerful subject positions giving up some agenda-setting power. People living in Potchefstroom cannot drive collective agendas across race class, geography and other lines of division, when they are clearly a minority within the larger city. Here the iterative process between symbolic orders might include 'third', 'fourth', 'fifth' etc., of course non-linear 'movements' towards common vocabularies and signifiers and signified objects of collective concern. For reasons explained in Part I of this book, the latter objects of collective concern include the multifaceted problem of crime.

Conclusion

This chapter has analysed the discursive entanglements of crime reporting in the *Potchefstroom Herald* as an important part of the echo chamber effect of crime discourse. By taking news reports in the *Potchefstroom Herald* as the unit of analysis, the chapter has augmented the consistent Laclauian analytics of this book with Lacanian psychoanalysis. Key themes that emerged by drawing on this approach has been the introduction of a 'double-movement' between (unbounded) symbolic orders, as both a simplification of what is actually a life-long iterative process, and as something not fully observable in the work of Lacan. The chapter has also engaged with the psychoanalytic notion of the death drive. These tools have offered yet another way of distinguishing within what some might view as a relatively homogenous group – white Afrikaans speaking residents of Potchefstroom. As with other chapters, there is a clear purpose in pointing out these differences. They allow points of entry, of leverage through which, in this case, movement between symbolic orders or transcending particular approximate symbolic orders might be helpful in pointing towards potential floating signifiers and subsequent progressive hegemonic rearticulations. Signifiers such as 'barbaric', 'Afrikaner' and 'criminal' are all problematic for different reasons, be it because of the harm inherent in the words and/or the essentially sutured nature of these signifiers. Changes between symbolic orders, or more precisely understood, as continued resocialisation, should result in greater and more productive interaction between historical lines of division. Sincere interaction in this regard will likely require material sacrifices by elites, but, in the spirit of such interaction, the sacrifices should ideally be voluntary, at least to an extent. To this end continuous and active rearticulation of the social is needed.

The chapter also highlighted signs of hope, amidst the trauma of the Real evidenced in the content of the weekly newspaper. These include, a column explicitly aimed at more positive and constructive discourse, as well as empathy and solidarity amongst some more historically privileged members of society with some residents of Ikageng. Although it still appears to be a minority view, there are many instances of praise for the SAPS and other public servants from members of the economic elite. Likely champions to overcoming lines of division are these members of society, already inhabiting more progressive symbolic orders and it will depend on their ability to find partners and upscale articulatory processes. This will also likely require giving up agenda setting power, especially where interventions into the root causes of crime are planned. In such instances partners, who know their local contexts, are more likely to understand what is required. Even then, interventions should be planned with different time horizons. To revisit the vignette at the beginning of this chapter, we may conclude that the trauma of the Real. In other words, the extra-linguistic/extra-symbolic affect experienced by consumers of the *Potchefstroom Herald*, are unlikely to dissipate, without interventions into the root causes of crime, and the unproductive attitudes of some residents. These unproductive attitudes currently form part of the ontological structures which shape the laager.

Note

1   I should reiterate a comment made in a previous chapter. There is no 'correct' level of fear. My intension is not to tell people their fear is irrational. Instead my concern is with the ideas and practices that tend to latch onto crime and the fear thereof. These ideas and practices are often quite divisive.

References

Abboud, S., Dahi, O.S., Hazbun, W., Grove, N.S., Hindawi, C.P., Mouawad, J. and Hermez, S. 2018). Towards a Beirut School of critical security studies. *Critical Studies on Security*, 6(3), pp. 273–295.

Adams, R. 2018. Pukke debatteer kampuskultuur. *Potchefstroom Herald*. 18 July, p. 3.

Anon. 2019. Redes waarop Potchefstroom agteruitgaan. *Potchefstroom Herald*. 9 June, p. 6.

Anon. 2018a. Geen geld vir veiligheid. *Potchefstroom Herald.* 3 May, p. 4.

Anon. 2018b. Commentary. *Potchefstroom Herald.* 17 May, p. 6.

Breitenbach, D. 2020. Newspapers ABC Q4 2019: A lacklustre final quarter for newspapers. Online: www.bizcommunity.com/Article/196/90/200639.html. Date of access: 10 May 2020.

Cho, D. 2006. Thanatos and civilization: Lacan, Marcuse, and the death drive. *Policy Futures in Education,* 4(1), pp. 18–30.

Cilliers, J. 2019. Potch dis tyd vir opruim! *Potchefstroom Herald.* 18 April, p. 11.

CrimestatsSA, 2019. Ikageng. Online: www.crimstatssa.com. Date of Access: 20 January 2020.

Fanon, F. 2004 edition. *The wretched of the earth.* New York: Grove Press.

Farmer, P. 2001. An anthropology of structural violence. *Current Anthropology,* 45(3), pp. 305–325.

Feldman, S. 1987. *Jacques Lacan and the adventure of insight. Psychoanalysis in contemporary culture.* Cambridge, MA: Harvard University Press.

Galtung, J. 1969. Violence, peace and security. *Journal of Peace Research,* 27(3), pp. 167–191.

Herman, D. 2016. Universiteit is meer as net 'n taal. *Maroela Media.* Online: https://maroelamedia.co.za/debat/meningsvormers/universiteit-is-meer-as-taal/ Date of access: 19 May 2020.

Homer, S. 2005. *Jacques Lacan.* London: Routledge.

JB Marks Municipality. 2018. Annual report. Online: https://jbmarks.co.za/sites/default/files/2019-06%20documents/2017-2018%20Annual%20Report.pdf. Date of access: 25 May 2020.

Lacan, J. 1960–1961. *Seminar book VIII: Transference.* Translated by Cormac Gallagher from unedited French typescripts. Online: www.lacaninireland.com. Date of access: 28 July 2020.

Laclau, E. and Mouffe, C. 2014[1985]. *Hegemony and socialist strategy.* London: Verso.

Laurita, C. 2010. Working with the drive: A Lacanian psychoanalytic approach to the treatment of addictions. Doctoral dissertation, Duquesne University. Online: https://dsc.duq.edu/etd/805. Date of access: 20 May 2020.

Leshage, S. 2018. Municipality to rezone parks to create housing development. *Potchefstroom Herald.* 9 August, p. 4.

Luxenberg, T., Spinazzola, J., Hidalgo, J., Hunt, C. and van der Kolk, B.A. 2001b. Complex trauma and disorders of extreme stress (DESNOS) diagnosis, part two: Treatment. *Directions in Psychiatry,* 21. Lesson 26, pp. 395–415.

Luxenberg, T., Spinazzola, J. and van der Kolk, B.A. 2001a. Complex trauma and disorders of extreme stress (DESNOS) diagnosis, part one: Assessment. *Directions in Psychiatry,* 21. Lesson 25, pp. 375–392.

Maphanga, C. 2020. Sanco demands suspension of JB Marks mayor and municipal manager. *News24.* Online: www.news24.com/SouthAfrica/News/sanco-demands-suspension-of-jb-marks-mayor-and-municipal-manager-20200127. Date of Access: 19 May 2020.

Mulder, C. 2018. 'n Enorme Afrikaanse 'nee' vir NWU se taalplan. *Potchefstroom Herald,* 5 April, p. 8.

Oosthuizen, J. 2019. Die tameletjie van veeltalige universiteite. *Litnet.* Online: www.litnet.co.za/die-tameletjie-van-veeltalige-universiteite/ Date of access: 19 May 2020.

Scholtz, L. and Scholtz, I. 2008. Die debat oor die posisie van Afrikaans aan die Universiteit Stellenbosch:'n Ontleding. *Tydskrif vir Geesteswetenskappe,* 48(3), pp. 292–313.

Smalman, I. 2019. Moenie frustrasie op die polisie projekteer nie. *Potchefstroom Herald.* 4 July, p. 6.

Van Riet, G. 2020. Intermediating between conflict and security: Private security companies as infrastructures of security in post-apartheid South Africa. *Politikon: South African Journal of Political Studies*, 47(1), pp. 81–98.

Van Riet, G. 2017. *The institutionalisation of disaster risk reduction: South Africa and neoliberal governmentality.* London: Routledge.

Van Riet, G. 2016. The limits of constitutionalism and political development. *New Contree*, 75, pp. 98–115.

Vincent, B. 2020. Jouissance and death drive in Lacan's teaching. *Ágora: Estudos em Teoria Psicanalítica*, XXIII, pp. 49–56.

Wetdewich, D. 2018. Station commander calls for crime summit. *Potchefstroom Herald.* 17 May, p. 4.

Interview

Potchefstroom Herald Journalist. Interviewed by Gideon van Riet. 27 June 2017. Potchefstroom.

6 Private security companies and securitisation

Introduction

Private security companies (PSCs) have become an infrastructure through which assumptions of what are legitimate security concerns and how security should be provided circulates. As such, they are worthy of study in terms of their contribution to securitisation. Previous chapters, from Parts I and II, have demonstrated that such ideas do indeed circulate. To reiterate the brief mention of the concept in Chapter 2, by securitisation I simply mean the rendering of an object or process as a matter of and for security, to a large extent to be performed by these PSCs. Although, in each case discussed immediately in the following, logics of equivalence typically frame the situation as one group in need of protection from another. As per securitisation theories, either a referent object in need of protection from a perceived threat can be securitised or a potential threat to a general population can be securitised. Such securitisation can take on at least two forms. Securitisation, when successful, may invoke an emergency politics that justifies the suspension of deliberation on the matter. The primary objective becomes mobilising resources for security. Alternatively, securitisation might be associated with routine practices that are still intrusive to some, but less overtly framed as a type of acute crisis (cf. Balzacq et al., 2016; Balzacq, 2011, 2005; Buzan et al., 1998). The former perspective is known as the so-called Copenhagen School and the latter as the Paris School of security analysis. This chapter is more concerned with threat construction through articulation, as in the Copenhagen approach. The chapter demonstrates the presence of securitising moves at the level of the town of Potchefstroom. The broad circulation of these ideas, and the fact that PSCs and other security actors often act upon these ideas, suggests that securitisation, as emergency construction, is often successful. The three chapters included in Part III that follows this chapter is more interested in the routine practices associated with a PARIS approach. Therefore, emergency construction is not all that is at play regarding crime in Potchefstroom. The picture is more complex. A recurrent theme throughout this book is how social and other media serve as security infrastructures pertaining to crime and the response to crime. Notions of securitisation have been implied in previous chapters. This chapter makes securitisation explicit and locates it within the broader echo chamber effect.

DOI: 10.4324/9781003028185-9

Of the many ways how PSCs have been analysed in recent years little, if any, attention has been given to their social media platforms and how they interact at a discursive level with other media platforms. This chapter analyses the public Facebook pages of PSCs to understand the type of securitisation mediated through this platform. The sheer volume of participants and followers of these pages make them significant in regards to securitisation and their involvement in the echo chamber that amplifies ideas around crime and security. Therefore, PSCs, much like the CPF and the *Potchefstroom Herald*, participate actively and passively in securitisation as they provide content for these pages themselves, whilst these platforms also allow for content to be added by the public and other infrastructures, such as the local newspaper, the CPF and the CID. The public served by these companies and to which these pages are primarily meant to speak, is largely white, middleclass and very religious. This does have a bearing on the content shared by the PSCs and especially the public.

I have chosen the two largest PSCs in Potchefstroom, because their social media presence is the most significant and they share the overwhelming majority of the market. I deduced this fact from observing the number of vehicles each company has on the road and by asking armed responders and managers to estimate the relative size of the different companies. Both companies add new content numerous times each month. They tend to focus on Facebook as their main social media presence. Their Twitter presence is miniscule and therefore of little concern here. I will, however, argue that their Facebook presence is significant for a number of reasons. They serve the echo chamber effect by reiterating and adding unfavourable connotations to the problem of crime. These security companies, wittingly or unwittingly, help to do this in a number of ways. Most significantly, they provide yet another platform for the public to speak and reinforce particular insular and exclusionary logics, most of which can be collapsed into the metaphor of the laager. One of these companies in particular, often shares rather militaristic ideas of security. Moreover, these social media platforms also allow for what we might call 'piggy-back marketing'. Obviously the Facebook pages exist primarily to serve a marketing function for the PSCs themselves, even though this might be cloaked in notions of community service. In addition, other organisations or groups also use PSC Facebook pages as marketing platforms. This is different from so-called 'ambush marketing'. The latter implies that the primary organisation providing the platform does not consent to the content being shared. This cannot be the case for PSCs, as administrators of Facebook pages can and do delete content. It is likely that some filtering does take place and that these companies do indeed practise a fair degree of social responsibility regarding the content they allow on their social media platforms. I would however argue that more can be done.

In terms of research ethics this chapter presented a slight trade off. As with references to the Potchefstroom CPF Facebook page, no names are mentioned, when quotations are cited to illustrate a particular point. As with the CPF group the date is still given, for the sake of transparency. The analysis of PSC Facebook pages created an additional ethical consideration. For the sake of transparency, a reference to the PSC site and the date has been provided, where quotations are cited. I therefore I had to name the PSC whose site the quotations have been taken from. As I will explain, however, especially the most problematic content is not added by the PSCs themselves, but rather by the public. The potential loss in privacy (revealing the name of a particular PSC) can be ethically justified if we weigh up the potentially damaging nature of discourses that circulate through these and other platforms against the naming of a particular platform. Moreover, both these Facebook pages are public and critical assessment of their content by outsiders cannot be viewed as unreasonable or completely unforeseeable.

In this chapter I analyse posts from January 2017 until December 2019 on the Facebook pages of the aforementioned PSCs. According to Facebook 8004 people liked and 8,162 followed the one PSC's page and 8,825 people liked and 8,928 people followed the second PSC's page on 22 January 2020. Before providing the outline of this chapter, I must introduce one more concept pertinent to this chapter. Giddens's (1990) concept of ontological security has been used in security to security studies (cf. Krahman, 2018). The concept is also of interest here. Ontological security refers to a sense of security or feeling of security. It relates to one's position in the world and a type comfort and certainty around this position. The world makes sense, and it is likely to do so for the foreseeable future. Krahman (2018) uses a lack of ontological security to explain why even though most of Europe is safer today than ever before, many feel insecure. She argues that by being able to analyse 'unknown unknowns' through risk assessment and management, PSCs offer many residents of the EU increased ontological security. I would argue that aspects such as language rights (discussed in Chapter 5), crime and security and fears about land expropriation often causes a great deal of ontological insecurity for a great number of the residents of Potchefstroom. While PSCs may indeed increase ontological security for many residents, the circulation of ideas through security infrastructures and the echo chamber effect demonstrated in this part of the book, also decreases ontological security for many. Therefore, crime and ancillary discourses that circulate through PSC Facebook pages, can be rather problematic. I have reiterated throughout this section of the book that I am not making an argument against the reality of crime. My argument is against the idea that 'they are out to get us'. Again the ubiquitous 'they', should be understood as a composite based on a logic of equivalence, stemming from the radical contingency of the social. There are other ways to see and compose the social. If there is hope for any meaningful post-apartheid South African project, incremental steps towards partnership and hopefully complex fusions between contemporary sutured notions of 'us' and 'them' will be required.

The chapter is structured as follows. The next section provides a general overview of the types of content frequently shared on PSC Facebook pages. I will offer a critique in subsequent sections, mostly to the responses that this content elicits. I will elaborate the argument that these platforms, although serving some meaningful security purposes, are also, probably mostly inadvertently, complicit in the securitisation of the laager and associated insular practices. Particular projected spaces of representation following Lefebvre (1991), aligned with the contemporary sutured hegemonic order dominate in the discourse that characterises these platforms. However, the chapter concludes by searching for signs of hope that may lead to progressive change, as opposed to the conservatism and even reactionary change promoted by some residents of the city. These signs of promise are found in the social capital that already exists for example in the collaboration by security actors in the city. What is required, is greater emphasis on some ideas, such as the rights of vagrants, and that we point out the floating nature, and as such the inherent potential, of some problematic signifiers. These overlapping signifiers include, race, moral decay, ontological security and the local economy. They may be deconstructed and re-sutured as a series of collaborative democratic practices.

The content shared by PSCs on their Facebook pages

For the purposes of marketing, PSCs use Facebook pages to show off the latest technology they have available to customers. At the time of writing the latest technology was CCTV cameras monitored live around the clock. This technology is a major focus of later chapters dealing with the Oewersig neighbourhood residents' association (Chapter 8) and the newly formed Cachet Park CID (Chapter 9). Suffice it to say here that fibre technology is now available in much of the city. This makes is it economically and practically feasible to utilise CCTV cameras for live monitoring of streets. Both companies focussed on in this chapter provide this service. Other technologies such as electric fences, laser beams, that trigger alarms when disturbed, and alarm systems are also showed off through advertisements and special offers or through incident reports. Security companies often reported on their timelines when they arrested someone.

Security companies often post, what we may call public service announcements on their Facebook pages. These posts might vary from simple lost and found information and missing persons reports, to giving thanks to loyal clients, to more or less overt marketing. One of the PSCs, for example, offers brief monthly crime statistics for cases they are aware of (they were called to). These statistics are shared at the level of the entire city, while they also provide a link to statistics broken down for each suburb or part of the city. Safety tips are often combined with an advertisement for a product sold by the company. One example would be every year during November when PSCs encourage residents to reach out to neighbours in light of the coming summer holidays. Knowing your neighbours would increase the odds of them 'keeping an eye' on your property when you are away. The PSC might follow up on this message at the beginning of December with a link to a form that clients can complete if they are going away during the December holidays. The PSC would then be aware of this fact and pay special attention to the property. Public service announcements such as these are related to overt attempts to be known as a 'community organisation'. Both companies to some extent try to market themselves in this way, but the one is far more focussed on this approach to marketing than the other.

Serving the community is even part of that company's slogan. The second and smaller PSC is more militaristic in its messaging. It would often post pictures of military style training that employees underwent. On other occasions public service meets overt marketing more directly. One of the companies developed an application that can be downloaded onto a smartphone. This essentially turns the phone into a georeferenced panic button. An advert for this app that clients can subscribe to would be shared and combined with messages stating that it is unsafe to jog alone in the street or to drive home alone in traffic. The successful apprehension of criminals is a second major part of the content shared on PSC social media platforms. These posts often contain within them notions of what constitutes 'suspicious activity' and areas under threat. Thereby PSCs securitise certain forms of behaviour as threatening and certain places referent objects. In a sense race and vagrancy are securitised. I will be elaborate on this form of securitisation in the following.

We might label another category of posts by PSCs as credentials sharing. These include not only the Private Security Industry Regulating Authority (PSIRA)[1] certification that is more widely shared, but also in specific instances, where staff graduate with certifications, for example on alarm installation. Security companies might also share information of training they underwent or training they gave to clients. Some of the training offered demonstrate links with problematic characters. For example, one form of anti-rape training offered and documented on Facebook is based on a self-defence system, based on interpersonal combat techniques, developed by a former Israeli special forces operative. He is on record as stating that farm attacks are acts of terrorism (Gous, 2017), a comment shared through the crime echo chamber. This type of language arguably fuels the contemporary notions of an onslaught by mysterious insurgents that threaten 'civilised' life, described in previous chapters. Teaching self-defence is a worthy enterprise. Teaching women to protect themselves in a country with high instances of rape, is even more laudable. In no way is the aforementioned critique against any of these two worthy causes. The question should however be asked whether it is wise to link the problem of self-defence to terrorism, when the link is dubious.

On other occasions, PSCs leverage the religious nature of the target population. One of the companies in particular would often share a 'Bible verse for the day'. They also posted when they were tasked with protecting a prominent lay preacher, who evoked widespread discontent for stating that only Jewish and Afrikaans people have a covenant with God (Rall, 2019). Yet, he also has significant support under conservative segments of the population nationally.

A critique of contemporary securitisations

It is arguably understandable that PSCs use social media for marketing purposes. Moreover, some of the security tips they give and content they share are likely very useful to the general public. Knowing when and where particular types of crime tend occur might, for example, inform planning by the public and as a consequence some many might feel safer and even be safer. They might there-fore, indeed as Krahman (2018) notes experience improved ontological security in addition to physical security. Therefore, this chapter should not be viewed as a wholesale critique of PSC social media sites. Having noted this very important qualification, there are matters the state, the public, PSCs and other concerned parties should consider. These mostly relate to the responses by the public added to the content shared by PSCs. In other words, my concern is mostly with the content added by the public and other organisations to these sites. It is oftentimes this content that is more likely to undermine ontological security.

These comments can be clustered around a number of themes. The first of these themes are disparaging remarks about the criminal justice system, including the SAPS, as well as the state in general. While many of these comments have some merit, they fall far short of any notion of 'finding solutions' to key problems in society. They are also ignorant, wilful or not, of the democratic values and human rights the post-apartheid criminal justice system is meant to observe. At the same time PSCs are praised following almost every piece of content shared. One might even speculate that this is an intended consequence of at least some of the content provided. In other words, praise is elicited from clients and those who follow these pages. Clearly the public trusts PSCs more than they do the SAPS and the state, who is at least partially deemed to be an opponent. This brand of discourse, which unfavourably compares the state with PSCs, seemingly reinforces notions of onslaught and uncertainty about the future. It should be acknowledged that a small minority of commentators, probably far less than the actual clients of these PSCs are members of the newer political elite. Their comments are far fewer, and far less significant in the circulation of these ideas of onslaught that promote ontological insecurity. This is probably because, they are more comfortable as regards their position in the post-apartheid social order. The following comment is indicative of a far more frequently observed trend. A PSC would typically explain how a suspect was apprehended and handed over to the SAPS, as is required by sections 39 and 40 and Schedule I of the *Criminal Procedure Act* (South Africa, 1977). Sharing information of these essentially citizens' arrests frequently lead to comments such as the following.

> The criminal probably got bail.
>
> (MB Facebook group, 14 May 2019.
> Author's translation)

It appears that some members of the public are ignorant of the due process that cases have to go through. Suspects often qualify for bail, unless the crime is extremely serious or the suspect poses a significant flight risk. The fact that the criminal justice system is overloaded, including prisons, the proper administration and investigation procedures required of the SAPS and the case load of prosecu-tors, makes release on bail or the non-prosecution of flimsy cases even more likely. In these instances, both the criminal justice system as a potential threat is securi-tised as well as a particular undefined, though thinly veiled, white, religious and

Afrikaans community. The former is the threat. The latter is the referent object, in need of protection. Other demonstrative comments include the following:

> (We) have no more protection in South Africa anymore.
>
> (CB Facebook page, 1 April 2019)

> What has happened to *my* right to freedom.
>
> (CB Facebook group, 16 January 2019,
> author's translation, emphasis added)

The second problematic discourse is the dehumanisation of suspected criminals. Frequently, suspects or criminals are referred to as 'rubbish' or 'things' (MB Facebook group, 11 January 2017). We should also view these comments in conjunction with the blurred photos often shared arrests. The photos typically display restrained black men. The result, at times, appears as a toxic reiteration of apartheid era images of the black insurgent other. This time he is part of the new dispensation that some perceive to be out to get them. Within this context the black vagrant is often pitted against the white property owner who is employed, just like state police and black politicians are pitted against private security. In other words, race is securitised as a threat to the life, limb and the property of an in group, which seems to be based on an equivalence of race, social status and language. Words such as 'things' and 'rubbish' also bring back into view Mary Douglas's notion of dirt as matter out of place (Douglas, 1966:44). The supposed purity (of a sutured) in group is juxtaposed to dangerous outsiders.

A third problematic theme that emerges from some comments by members of the public is a type of nostalgia for the autocratic past. These have been accompanied by memes that display the apartheid era South African flag and the four-colour flag of the ZAR, the old Afrikaner republic covering in today's terms the Limpopo, Mpumalanga, Gauteng and most of the North-West provinces. This republic existed until British colonisation of the whole of modern-day South Africa subsequent to the South African War of 1899–1902. Importantly, this dehumanisation of others described earlier is frequently linked to a narrative of a society in decay. I have in previous chapters made an argument for Strain Theory as a major explanatory tool for crime in South Africa, which is linked to the notion of anomie. We should, however, remember that anomie *is not* a state of normlessness. It is rather a state of norm confusion. One might even associate anomie with a lack of ontological security, which implicates the society as a whole and not just criminals. Therefore, 'decay' is not a function of the criminal or something to be projected onto particular segments of the social order only. It is also a function of the social order in general, of which the complainant too is a part. Moreover, 'decay' as a discourse that harks back to a supposedly morally correct apartheid system makes little sense. It is also ethically unjustifiable. Autocracy was not a more moral era. It simply displayed different hegemonic norms, many of which were extremely cruel. At its very core was the norm that some people are less human than others. References to Potchefstroom as the 'Wild West' and categories such as '*ongewenstes*' (the unwanted) are further indicators of this mentality (MB Facebook group, 11 September 2019). The 'Wild West' may imply ontological insecurity as people feel extremely unsafe in a perceived state of disorder. It is a black run government and city council that is unable to provide such order. 'The unwanted' clearly refers to a category of would-be criminals, likely linked to the suturing of race, class and crime. It speaks to a type of insularity aimed at excluding an unwanted, poorly defined other. A more legitimate security concern would have been articulated in more specific and less prejudicial terms. To this we should add articulations such as 'evil', 'revolution' and 'vengeance' linked to a particular post on one of the PSC Facebook pages. These concepts speak to a perceived deviousness and extreme ontological insecurity in light of a potentially dramatic and radical restructuring of society in which those who posted the comments may have quite a lot to lose. 'Vengeance' might even suggest projected guilt. In this specific instance, the fears evoked were potentially incited by content added by one of the PSCs where a picture was shared with the tagline 'facing off with the EFF' (CB Facebook group, 21 June 2018, Author's translation). This picture was captured during a protest. The EFF, are the Economic Freedom Fighters, a socialist party that split from the ANC in 2013. They have been avid promotors of land reform. Sharing such content therefore clearly plays into the fear of many residents and followers of the particular Facebook page.

A fourth set of potentially problematic themes stem from the 'piggy-back marketing' mentioned previously. One of the PSCs are intimately involved in the newly formed Cachet Park CID. It provides a designated armed response vehicle for that area. It provides security guards that are stationed in the area and it monitors around numerous cameras installed in the CID. This PSCs Facebook page has recently become a platform for comments from partners in the CID. The PSC in question also adds content of its own promoting the CID. The CID is the subject matter of an entire chapter, so I do not want to go into the details of how it works and why it was formed. The point here pertains to the messages that have been shared on the PSC page regarding the CID. These include taking back 'our place', 'cleanliness', and there is 'no more place for loiterers' and 'car guards' as the PSC now fulfils that function. In the process vagrants and car guards are securitised as threats to local business, car owners and the general public.

In addition, the securitisation of businesses as referent objects in this part of the city, because criminals drive customers away, allows to flourish the discourses of 'us' as benevolent by giving 'them' jobs in 'our' spaces. Crime threatens this type of dissemination of wealth, as the subtext in the following quotation taken from the CPF group attests:

> I see they are also busy on the Bult? So, are we going to avoid all those places (businesses) so that they get their way and other people lose their jobs.
>
> (JB Marks Security Facebook group, 19 January 2018,
> Author's translation)

This type of comment is ignorant of the terms by which such employment takes place and the lack of employment opportunities in the country as explained in Chapter 1. Furthermore, this massive unemployment contributes to South Africa's status as one of the most unequal societies in the world. In this context legiti- mate security concerns have morphed into a structure that is violent that perpetu- ates the widespread structural violence experienced throughout South Africa (see Van Riet, 2017). 'Loitering' is also securitised in a country where freedom of movement is enshrined in Section 21 of *The Constitution of the Republic of South Africa* (South Africa, 1996). Also consider the following, a statement that all but criminalises, never mind securitises, vagrancy:

> We experience the difference that (name of PSC) makes in *Oewersig*. The guards do a good job. It can only work positively on the *Bult*. It takes a while before these scoundrels realise that there is no more opportunity for them in the streets. But, in the long run, when they all get the message, it will work.
>
> (MB Facebook group, 1 May 2019,
> Author's translation)

What this instance also indicates is the gulf between South Africans, that is largely enshrined through infrastructure. The old apartheid city planning of middleclass suburbs, 'shielded' the white and wealthier population from more impoverished black settlements by industrial areas or open spaces has changed only partially after apartheid. Since the end of apartheid, the state has built many homes for homeless South Africans. Notwithstanding the changes in structure of the con- temporary hegemonic order, these developments have largely observed much of the old spatial template. This arguably makes ignorance and cruelty possible. Infrastructures such as the CID, meant to protect students and business from crime also have the potential to do the same, especially if vagrancy, which is a major reality in South Africa, affecting a significant proportion of society, is securitised as a threat. What these residents do not understand is that many vagrants choose to live and sleep inside the proverbial laager as it is safer and a place where a liv- ing can be scavenged for example through collecting recyclable materials or car guarding.

Afriforum has also piggy-backed its own promotion on the platform of PSC Facebook pages. This organisation brands itself as a civil society organisation and have been said to be playing on the fear of a 'white genocide' (Chothia, 2018). As previously noted, it has also been accused of primarily advocating for white minority interests South Africa (cf. Pogue, 2019; Associated Press, 2018; Basson, 2011). As such, they reinforce ontological insecurity amongst those who feel the status quo, contra the old order, is hostile towards them. In 2018 their leaders even had an audience with Fox News in USA, where they promoted concerns over land expropriation. The organisation synergises their neighbourhood watches in Potch- efstroom with one of PSC's patrols. Both patrols have contact through shared radios. I was once privy to an interaction during a night shift, when I accompanied a PSC on patrol. We were called by the Afriforum vehicle the driver who saw black children walking in the streets and standing in front of a petrol station. The Afriforum driver's concern was literally expressed as follows: 'What are they doing on a Friday night in front of a petrol station'. Seemingly, at least some members of this organisation are still harking back to the romance of an exclusively white part of the city. Therefore, association with the organisation is likely ill-advised in a democratic setting. Still, invites to join Afriforum can be found on the pages of these PSCs. Instances of apprehension and arrest of suspects in collaboration with Afriforum are also shared on PSC Facebook pages. My concern here is with the overt linkages with, and support for, a clearly divisive organisation, regardless of their actual official intentions. Knowledge of such affiliations are likely to make PSCs (even more) unpopular with the majority of residents of Ikageng who often come into Potchefstroom for work. Such divisions are not the stuff of nation-building or in Laclauian terms progressive rearticulation towards radical and plural democracy.

In another instance of piggy-back marketing, the university's own rental agency distinguishes itself from others by adding security related value to sites it rents out. Consider the following:

> IceBear and Pukki-Verblyf Student Accommodation took hands to improve the safety of students who rent from the agency. We are proud of the difference we are going to make. We will keep you up to speed!
>
> (MB Facebook group, August 10, 2019.
> Author's translation)

Icebear is the app that has been developed by the one PSC that allows a smartphone to be turned into a georeferenced panic button. At first glance this post is completely unproblematic, and to a large extent it is. There is arguably no intent to harm anyone. Icebear is however significant as it affirms the images conveyed by adverts for this app. That the app is useful and ingenious is clear. However, stating that one should not jog alone without the app, reinforces notions of widespread panic and ontological insecurity, when the reality is that many people do go for runs every day without suffering any harm. Therefore, the point I am making is one pertaining to the circulation of ideas and connotations linked to safety. In this instance, an arguably ill-advised connotation linked to a generally legitimate concern and potentially valuable piece of technology with security, is unwittingly reaffirmed by a third party.

All the themes raised in this section links to a fifth theme. There is a tendency in the response to crime to emphasise infrastructure and environmental design, while all but ignoring the deeply seated socio-political aspects of crime. This response to crime translates into pushing back the laager deeper and deeper into the proverbial frontier. In the process crime becomes someone else's problem, for the most part. Many of the fundamental causes, however, are not addressed, possibly because they are far more difficult to do and to a large extent dependent on a wider array of expertise than security services. These concerns are highlighted by citizens who voice their concern that crime will simply be directed elsewhere with the increased security infrastructures in one part of the city.

> Will the criminals just move to another part of the town?
>
> (MB Facebook group, 1 May 2019,
> Author's translation)

This concern is consistent with the literature on security that have posited and sometimes found that such displacement of crime does take place when interventions are launched in a particular area (Krahman, 2008:132; Sandler, 2001). Infrastructure and environmental design do nothing for inequality and abject poverty. They are not designed to do so. Yet, the fundamental political-economic challenges that the South African polity faces continues, as is evidenced by its incredibly high unemployment rate and its continual ranking amongst the most materially unequal societies in the word. The evidence suggests that security infrastructures in Potchefstroom, as promoted on the Facebook pages of PSCs, have sometimes wittingly and/or unwittingly exacerbated these root causes of crime by deepening the gulf between those with and those without means.

Signs of promise and potential praxes

In light of the sutured nature of the social, often based on logics of equivalence I will highlight floating signifiers and practices in this section, that potentiate disruption. These signifiers can be challenged and rearticulated through identifying points of rupture or dislocation in the radically contingent social order. It is perhaps worth reiterating here what Laclau and Mouffe (2014[1985]) mean by radical democracy. Democracy is a word that has many connotations and I must be very clear on what I mean and why I am concerned with certain undemocratic sentiments that circulate in Potchefstroom. By democratic sentiments, I mean a prevailing attitude of hearing and taking seriously the voices and interests of diverse segments of the population. Thus, radical democracy entails the pluralisation of struggles and instances of mutual solidarity. This extends far beyond elections. It pertains to, 'a new common sense' that is not based on a supposedly completely sutured or closed set of competing interests (Laclau and Mouffe, 2014[1985]:167). There is always room for constructive collaborations, as the social is always in flux. Logics of equivalence and the dynamic categories they render are recognised for exactly what they are, based on radical contingency and the over determination of the social. Clearly, much of what has been described earlier does not conform to this definition. I have identified signifiers throughout this chapter that might provide points of entry en route to a more democratic politics. There has been a general reiteration of the laager, as a collective infrastructure, through the diverse securitisations stated throughout this chapter. These include race, class, vagrancy, car guards, the moral order and the local economy and businesses.

Race remains associated with the exotic yet devious other, who for example employs 'African signs' to mark targets. Race is also linked to the securitisation of vagrancy and car guards, as most of these people are black. Comments to this effect cannot be allowed on a PSC social media platform. This change in discourse should not cause an increase in crime. The same monitoring of cameras, guarding and patrolling of areas will likely continue. Yet, harassing people for conforming to what Diphoorn (2015) calls the 'Bravo-Mike Syndrome' is very problematic given the realities of contemporary South Africa. This 'syndrome', following the NATO alphabet, implies the securitisation and rendering as suspicious of black men (Bravo Mikes) in historically white and middleclass parts of the city.

Another problematic signifier is the notion of moral decay. In addition to the aforementioned reference, such decay has been linked to increased drug dealing and addiction, throughout the city inside and outside of the laager. Here discerning the problematic from the unproblematic is relatively easy. The drug problem is clearly real and pervasive. There can be little doubt that intervention on a large scale is required. Using the signifier 'moral decay', however, has very little merit as a discourse in contemporary South Africa. The territory simply has too long of a history of questionable moral conduct by groups occupying dominant subject positions. This discourse has to be rearticulated as social capital and trust building, which will imply improving the relationship between those who fear and the general population to whom they attach the label of 'threat'. PSCs can play a major role by adjusting the discourse they allow on their Facebook pages. In part, this implies taking greater care with inadvertent racial profiling and some problematic alliances with divisive groups and organisations, that may promote ill-will towards the majority of citizens feared by a small yet significant group of citizens. The subsequent associations of PSCs with serving race- and class-based needs is a reality that needs to be changed. It perpetuates hate. Private security staff on various occasions confirmed or even brought up a sense of perceived hatred from a significant proportion of black residents.

Achieving this, that is, rendering PSC Facebook pages as active contributors to greater democratic sensibilities in society, should not be all that difficult. Collaboration already exists between a whole host of organisations in Potchefstroom for the sake of greater security. This includes PSCs, the university, local businesses and importantly the SAPS and the city council. Particular spaces of representation are, however, promoted through insular ideas and exclusionary practices. In the contemporary security echo chamber these ideas dominate over the wants of the majority of JB Marks residents. This might very well be driven by the economic power of local businesses. Therefore, we can impute 'local economic activity' as a presumed, and often articulated, signifier in the discussions that led to the current dispensation. This signifier needs to be dislocated by the city council, possibly with an argument that the local economy can be secured without impinging freedom of movement. Such moderate adjustments to the content of these pages might, in turn and in time, help change problematic attitudes of the audience and citizen contributors, and the connotations they link to the problem of crime.

Conclusion

This chapter has analysed the Facebook pages of the two largest PSCs in Potchefstroom. It has provided an overview of the content typically shared on these pages. It has also critiqued the content shared on these pages, often by third parties, whereby problematic securitising speech acts circulate. Primarily, it is the reiterative securitisation of the laager and its inhabitants that is framed as referent objects, threatened by unwanted others, that takes place. The chapter concluded by highlighting key floating signifiers that can be leveraged toward more democratic ends. These signifiers include race and vagrancy, moral decay and local economic activity. These signifiers help to sustain insular ideas that perpetuate the laager as a mentality and a physical and policed reality. The chapter has concluded that there should be greater 'push-back by the SAPS and the city council on the demands of other actors and greater filtering on the content posted on the Facebook pages of PSCs, to facilitate circulation of greater democratic sensibilities.

Note

1   PSIRA is the authority tasked with regulating the private security industry in South Africa. It is governed by the *Private Security Industry Regulation Act 56 of 2001* (South Africa, 2001). Every PSC must obtain PSIRA accreditation before it is considered legal. The body has inspectors who visit PSCs to establish their legality. They have the authority to arrest non-compliant company owners.

References

Associated Press. 2018. Pistorius prosecutor sets sights on South Africa's Zuma in Washington Post. Online: www.washingtonpost.com/world/africa/pistorius-prosecutor-sets-sights-on-south-africas-zuma/2018/02/18/811ebdbe-14b3-11e8-930c-45838ad0d77a_story.html. Date of access: 5 February 2020.

Balzacq, T. 2011. A theory of securitization: Origins, core assumptions, and variants, in T. Balzacq (ed.) *Securitization theory: How security problems emerge and dissolve*. London: Routledge. pp. 1–30.

Balzacq, T. 2005. The three faces of securitization: political agency, audience and context. *European Journal of International Relations*, 11(2), pp. 171–201.

Balzacq, T., Ruzicka, S. and Ruzicka, J. 2016. Securitization revisited: Theory and cases. *International Relations*, 30(4), pp. 494–531.

Basson, A. 2011. White first. African second. *News24*. Online: www.news24.com/Columnists/GuestColumn/White-first-African-second-20110926. Date of access: 5 February 2020.

Buzan, B, Wæver, O and de Wilde, J. 1998. *Security: A new framework for analysis*. Boulder: Lynne Rienner Publishers.

Chothia, F. 2018. South Africa: The groups playing on the fears of a 'white genocide'. *BBC News*. Online: www.bbc.com/news/world-africa-45336840. Date of access: 5 February 2020.

Diphoorn, T.G. 2015. The "Bravo Mike syndrome": Private security culture and racial profiling in South Africa. *Policing and Society*, 27(5), pp. 525–540.

Douglas, M. 1966. *Purity and danger*. London: Routledge.

Giddens, A. 1990. *The consequences of modernity.* London: Polity Press.

Gous, N. 2017. Farmers get military training to protect themselves against attacks. Online: www.dispatchlive.co.za/news/2017-08-28-farmers-get-military-training-to-protect-themselves-against-attacks/ Date of access: 3 August 2018.

Krahman, E. 2018. The market for ontological security. *European Security*, 27(3), pp. 356–373.

Krahman, E. 2008. Security: Collective good or commodity? *European Journal of International Relations*, 14(3), pp. 379–404.

Laclau, E. and Mouffe, C. 2014[1985]. *Hegemony and socialist strategy.* London: Verso.

Lefebvre, H. 1991. *The production of space.* Translated by Donald Nicholson-Smith. London: Blackwell.

Pogue, J. 2019. The myth of white genocide. Pullitzer Center. Online: https://pulitzercenter.org/reporting/myth-white-genocide. Date of access: 5 February 2020.

Rall, S. 2019. Fury persists over Buchan remark despite his apology. *IOL Online*: www.iol.co.za/mercury/news/fury-persists-over-buchan-remark-despite-his-apology-36873461. Date of access: 5 February 2020.

Sandler, T. 2001. *On financing global and international public goods.* Policy Research Working Paper 2638. Washington, DC: World Bank.

South Africa. 2001. *Commencement of the private security industry regulation act, 56 of 2001.* Pretoria: Government Printers.

South Africa. 1997. *Criminal procedure act 51 of 1977.* Pretoria: Government Printers.

South Africa. 1996. *The constitution of the Republic of South Africa. Act 108 of 1996.* Pretoria: Government Printers.

Van Riet, G. 2017. *The institutionalisation of disaster risk reduction: South Africa and neoliberal governmentality.* London: Routledge.

Part III

Everyday security practices

It is just after 01:00 on a Friday morning and I am woken up by the sound of tyres screeching followed by loud voices. A bulky young white man is beating a black man lying on the pavement about 150 m from my bedroom window. I live on the top floor of a block of flats and there is a very popular student watering hole to the South, while university residences and private student housing is mostly located to the North. I have seen some 'colourful' and despicable deeds since moving into this flat over the past seven or eight years. There are onlookers; car guards. I can tell this from the bibs they are wearing. The white man alternates between beating and pepper spraying the man on the ground. 'Wat is dit hier in die sak? Waar het jy dit gesteel. Hoekom breek jy by huise in?' (What is this in the bag? Where did you steal this? Why do you break into houses?) he shouts. The screeching tyres appear to have been from the white man's Toyota Hilux 4×4 pickup truck. It appears as if he chased the suspected burglar and then cornered him on the pavement. The black man is taking a serious beating. I want to get dressed and walk down the four stories of my flat block to the scene. This cannot go on. But, the incident, it seems at first, is soon over. The white man speeds away. The largest PSC in Potchefstroom and the SAPS arrive. This is unusual; that they arrive at the same time. 'The burglar' is lying on the grass, quite still. The white man returns with his pickup truck. He talks to the police. The PSC worker tells him 'go, just go'. The man on the grass screams. He is clearly in pain. The white man leaves again and an ambulance arrives. The scene clears just after 02:00.

*****

What is most concerning from this brief depiction is the entitlement the white man had in believing he was within his rights to torture a suspected burglar. The *Criminal Procedure Act 51 of 1977* is clear in stating that citizens may only apply appropriate force to protect themselves and restrain a suspect. This man was running away and hence not a threat. Moreover, what happened here clearly exceeded any notion of 'appropriate' and 'necessary' force in completing an arrest, be it a citizens' arrest or an arrest by the police. The white man apparently saw nothing wrong with what he did. He had no problem facing the police afterwards and they apparently let him go. We seem to have moved into a milieu where extra judicial and otherwise questionable responses to crime are increasingly acceptable.

DOI: 10.4324/9781003028185-10

Non-state security actors, including PSCs, semi-state institutions, the public and the institutions they develop, often operate in a grey zone between the public and the private spheres and, I would argue, between the legal and technically illegal. People in Potchefstroom are fed-up with crime. They are venting every day either in person and on social media. Although legally the white man's behaviour was criminal, it was not treated as such. This is very problematic.

In Potchefstroom, this sense of 'enough!' has as we have observed in previous chapters become attached to divisive discourses. Consequently, 'enough!' has translated into significant demands on the SAPS and PSCs and it has led to a set of daily security practices that foregrounds narrowly conceived spaces of representation in contrast to more inclusive representations of space. Meanwhile local government is all but bankrupt. Development at the local through the national spheres of government, to any meaningful extent, is a myth, as we have just experienced over a decade of directionless, leadership and even further back, an ill-conceived politics of austerity (see Chapter 1). The upshot has been the perpetuation of a hegemonic dynamic whereby security infrastructures, target hardening and environmental design reign supreme in maintaining the safety of political and economic elites. In Potchefstroom, the former often appears ambivalent or tepid in their response to how race is attached to narratives of crime. The laager as a safety blanket shrouded over relative privilege is reinforced and the paradoxical logics of 'closing' the frontier through biopolitical abandonment and frontier governance, through border-making endures.

This part of the book is comprised of three chapters. Chapter 7 provides an overview of the role and daily routines of PSCs, their relationship with the public and the SAPS. Chapter 8 investigates the operations of and the security measures instilled by the *Oewersig* Residents' association, in collaboration with the largest PSC in the city. Oewersig is an upper middleclass and as far as I am aware, a completely white, neighbourhood to the North of the city. Chapter 9 focusses on the Cachet Park CID formed in 2019, as a collaboration between the North-West University, the city council, local business owners and residents. In this particular context, the limited capacity of the SAPS paves the way for PSCs, residents' associations and partnerships, such as CIDs to perform frontier governance. When viewed in this context, the story relayed previously about the violent white man is not so much an aberration, as it is a particularly extreme example of a broader trend. Having noted this, exceptions, to various degrees, to such an exclusionary ontology, or the laager as an imaginary, are frequent.

7   Private security operatives

The typical and the stereotypical

Lisego used to work for the SAPS. He was an investigator, and then he worked in crime intelligence. But, he left the SAPS, because of 'human rights'. You see, criminals now have too many rights. 'You cannot do anything' or they will take you to court. I overhear co-workers talking to him about an event involving him at 'that' building, the one close to the Wandellaan where many criminals live. He chased a suspect into the building. When confronted with a locked door, he warned the occupants about what was going to happen if they did not open the door. Indeed, he did deliver on his promise. He fetched a crowbar from his car and broke into the flat.

In another instance, Lisego chased a cell phone thief into a drug house, where it was being sold. He jumped over a wall to enter the premises. The inhabitant then voluntarily gave up the phone, as long as Lisego left it at that. You do not look for trouble with Lisego. In this regard, Lisego is perhaps a stereotypical, but based on my observations in Potchefstroom, not necessarily a typical private security operative. There are various types of personalities working in private security. Many are quite even tempered, conscienes and reserved. Behaviour both repugnant and laudable can be quite context specific.

Later during the shift, we responded to a laptop being stolen from a bed shop in town. The manager expressed his feeling of helplessness. 'It serves no purpose to phone the police. They do nothing' the manager said. Another security worker whom I accompanied on a previous shift, had also made it to the scene. He replied to the shop manager that he (the shop manager) should shoot them. 'Then you just phone us to fix the scene before phoning the police'. 'Once you have (shot) one, they will not come back'. Maybe this was only a bad joke. In reference to my previous point on the context of good and bad behaviour, the latter respondent, when I accompanied him on patrol exhibited a profound humanity and a commitment to do as little harm as possible. He would not even carry a real gun. His weapon shoots pepper kernels and he is not the only one. There are different personalities within this industry. But, there are also discrepancies, even contradictions within a particular personality, which challenges common stereotypes. Moreover, this complexity might be instructive and useful for progressive praxes.

*****

DOI: 10.4324/9781003028185-11

Introduction

As mentioned in the brief account earlier, caricatures of armed responders exist. These stereotypes do fit in some instances. Equally true is the fact that stereotypes mask other characteristics. Armed responders do not universally align with the stereotype of hot headedness (cf. Van Tillburg, 2019; Smallhorne, 2017). Many, if not most working in Potchefstroom, come across as even tempered, consciences and careful. Others are less careful. This chapter analyses the role and nature of PSCs doing armed response in Potchefstroom. In the process the chapter will deal with the work these companies do and with the characteristics, behaviour and work conditions of the workers themselves. The latter is important, as it locates these individuals more comprehensively in the particular social order I am studying.

Security companies in Potchefstroom provide diverse security infrastructures, in the form of guards, including car guards, alarms triggered through sensors and panic buttons. They increasingly offer CCTV camera installations that are monitored live. They provide security at workers' strikes, VIP protection, the transport of valuable items, such as diamonds, and armed response to alarms. Not all of the PSCs in Potchefstroom do all of these, but the larger ones tend to provide most if not all of these services. Private security companies also interact with other infrastructures. They monitor social media and often respond based on concerns raised on WhatsApp groups. One company representative stated that they were linked to 38 different WhatsApp groups. If we were to exclude companies doing guarding only and include those that do armed response, then all the PSCs operating in Potchefstroom, bar one, are family-owned businesses. The exception is a local franchise of a South African company. Somewhat different to what has been reported in larger cities in South Africa and in other territories, no multinational companies do this kind of work in Potchefstroom (cf. Abrahamsen and Williams, 2010). Most, don't do patrols. They simply respond to alarms when triggered and they often only have one vehicle available for such responses. This chapter is concerned with the larger PSCs in the city that do patrols and, as such, are more significant in the politics of crime in the study area.

The analysis speaks to the commonly held notion that South Africa is a post-conflict society. Instead, South Africa may be viewed as a country not *at war*, but in *a state of war*, if one is to follow the Hobbesian distinction delineated by Foucault (2003:92). Whereas the former is fully blown violent conflict, the latter is part and parcel of the inherent conflict between groups that is never quite transcended and that is inherent to the definition of hegemony proposed by Laclau and Mouffe (2014[1985]). In this context PSCs intermediate between populations of different strata and in the process between conflict and security by protecting some, such as the political and economic elites and not others. They often do this by responding to and partaking in the circulation of the signifier 'suspicious' or 'suspect behaviour'. This represents much continuity with the role of the state police in pre-democratic South Africa. To stabilise the polity greater desegregation/integration of population groups is required (Van Riet, 2020). There is a need for constructive thought on this matter in order to, following Laclau (1990:8), not transcend the politics that cannot be transcended, but to rearticulate it continuously in increasingly less ominous forms. In other words, people have different and competing interests and therefore, politics is a normal way of life. What can be rearticulated is how this politics is practised.

In order to progressively rearticulate, the conflict between the political and economic elites and between elites and other parts of the population, we must understand the nature of the actors involved in potentially progressive change. The insistence on difference, not only between, but also within sutured categories and even the non-uniformity of individual actors, is a central thread that runs throughout this book. Private security operatives have become an increasingly important category of actor in the post-apartheid era and therefore warrant a dedicated chapter in this book. I wish to better understand armed responders and the companies they work for. I am not only interested in their 'front-stage behaviour' (Goffman, 1959), but also in the types of individuals, and complex assemblages of personalities that interact with other security infrastructures and clients. Once we have a better grasp of this, we may be sympathetic and critical at the same time and in the process hopefully offer some insights on how these actors may contribute or be drawn into the constructive rearticulation of politics in post-apartheid South Africa.

This chapter is primarily based on two types of sources. Firstly, I conducted interviews with four PSC owners and managers that operate in Potchefstroom. These interviews revealed a lot more than simply the nature of the business and their unique selling points. Owners and managers volunteered to speak about the different types of workers they have working for them and the difficulties each face. Secondly, I accompanied two PSCs on a number of shifts, more or less equally distributed between day shifts, night shifts, weekdays and weekends. These patrols were respectively with the largest and third largest PSCs in Potchefstroom, based on market share. It included ten different armed responders. Although this data cannot possibly speak to or challenge the conclusions of the voluminous literature on private security in South Africa and abroad, the data is saturated at the level of the study area to which this book speaks. Three companies share the overwhelming majority of the market and each of them only have a limited number of staff. During these patrols, I could more fully grasp the armed responders' orientations towards their work, their clients and the people they are meant to police and drawing on different armed responders from different companies served as a type of triangulation within the group of armed responders and with the interview data gathered from owners and managers. Both forms of data revealed a range of orientations towards the public, somewhat informed by different positionalities within the industry. In addition, the operational manager of the largest PSC in Potchefstroom completed a Masters of Business Administration thesis at the North-West University (NWU). I read through this thesis and made notes. It confirmed the large-scale staff turnover in the industry I had perceived and revealed some of the main causes of such turnover in the particular organisation. Finally, the analysis also draws on the social media analysis that largely informed parts of Part II of this book. This social media analysis added context in regards to the relationship between the PSCs and their clients.

The remainder of this chapter is structured as follows. The next section seeks to understand the nature of armed response work. The section will necessarily deal with remuneration, household background and trauma experienced on the job. Thereafter, I hone in on the diversity within the industry. I consider the often-asymmetrical relationship between clients and PSC owners and managers, and their employees and then between the PSCs and the SAPS. The chapter concludes by considering the sutured categories and floating signifiers that emerge from the analyses and it considers how diversity within this industry, read along with these sutured categories and floating signifiers, potentiates progressive change. It also reflects upon debates around hybrid and simultaneous authority in security provision (cf. Berg and Howell, 2017) and considers, given the analysis provided in this book, what principles 'hybridity' ought to be founded in.

By the end of this chapter 'the armed responder', as far as Potchefstroom is concerned, is demystified and engaged with as a malleable category. A second floating signifier is 'community' as a key marketing ploy of PSCs, so too is the notion of a 'force multiplier'. A fourth floating signifier is the notion of racism and racist agendas. This is perhaps controversial, as there is much incessant white on black racism that permeates South African society. Potchefstroom might even be worse than many other South African contexts. Often this racism circulates through PSCs. Therefore, to be absolutely clear, the analysis does not condone racism in any shape or form. My point is simply that, while it is structural throughout the social order, the extent and nature of internalised racism also varies between PSC operatives and positionalities. Finally, we may reconsider the signifier of the PSC as a security infrastructure constituted through power.

Being an armed responder

Private security is a key source of employment growth in South Africa's economy with incredible unemployment rates, as noted in Chapter 1. This fact was noted by Abrahamsen and Williams (2009:2) over a decade ago. The situation has not changed significantly. The demand for private security has continued to grow (Dzhingarov, 2017). Working in private security is, however, not always a preferred form of employment. Much of the evidence from Potchefstroom corresponds with the literature on private security in various parts of the world, which explains that this type of work is often dangerous, involves working long hours and receiving low wages (Chrisholm, 2016:180; Abrahamsen and Williams, 2010:178, 210, 2005:427–428).

Many armed responders feel trapped in their work for a host of reasons. South Africa has very high levels of unemployment. Armed responders are often paid per hour and salaries are clearly below that of even entry level SAPS officers, who according to an interview with a SAPS captain (Interview, August, 2018), earn around ZAR 13,000 per month. One manager voluntarily stated that the company he works for pays armed responders about ZAR 8,000 (USD 471) to ZAR

9,000 (USD 530) per month. However, various armed responders for example kept bringing up hypothetical salaries of ZAR 5,000 or ZAR 5,500 and that they would often move from one PSC to another for an increase of a few hundred Rand.[1] In terms of purchasing power these wage levels are significantly higher than what the directly converted amount would buy in the USA. However, it is still quite a low wage, especially given the dangers associated with this work. As a comparison, I know of receptionists at medical doctors' offices earning ZAR 15,000 per month at the same time. At the time of writing the national minimum wage in South Africa was ZAR 3,500. To further illustrate relative purchasing power, rent of a two-bedroom apartment, not in the wealthier North of the city, at the time was more or less ZAR 4,000–ZAR 5,000, while a white bread cost at least ZAR 12 and a two litre bottle of milk cost at least ZAR 25. Therefore, if the SAPS who pay higher salaries are understaffed and PSC, in a sense 'picks up the slack', then some of the costs of security in the middleclass suburbs are passed onto armed responders, who often live outside of the laager elaborated in Chapter 2. Some black armed responders live in Ikageng, in a new extension, which is still an informal settlement with few brick and mortar homes. One white armed responder lives in a budget student commune, inside another stigmatised space referred to in Chapter 3, specifically the part of the city with a high density of drug houses. His rent of ZAR 1,800 per month for a one room flat, excluding prepaid electricity, which would mean hundreds of Rand more each month, was lower than the overwhelming majority of student accommodations, closer to the university. Others mostly live in affordable flats in less affluent parts of Potchefstroom. Some of the latter are not the only breadwinners in their households.

Especially white armed responders tend towards low morale, largely linked to their work. For some black armed responders, who often grew up in informal settlements and whose parents would not have earned high salaries, armed response work might represent an improvement in living standard. They too, however, complained about money, just less frequently.

Armed responders come from a variety of backgrounds. One armed responder I encountered used to do anti-poaching work, in particular of rhinos, on a private game farm. Another used to be a technician installing alarm systems. Still others came from the public sector, having worked in the SAPS or Correctional Services (as prison wardens). Some previously had their own security companies. Others join the industry immediately after school or after working overseas for a year or two after school. Armed responders to not typically come from significant wealth. Often there is not enough money for further studies after school and the abundant PSC industry becomes a viable option. Owners of security companies and sometimes very senior managers are former military officers from the apartheid era or former members of the South African Police (SAP), the name of the apartheid era state police. Many of these older men left the military or police with democratisation and worked as mercenaries or as security consultants overseas. One current owner, for example, worked in Iraq guarding US military bases before buying and starting the provincial franchise of a particular PSC network. Over the past 25 years many people from the military, the police and correctional services left what they perceived to be unpleasant working conditions. This includes overfull and often violent prisons and dangerous and unappreciated work. Many believed that 'politics' meant that there was a ceiling to their prospects for promotion within the organisation. However, they may not have exchanged their previous jobs for what they had expected in terms of income and work conditions.

Having noted this historical pattern, I should also note that very few within the older generation are still in the system as armed responders. There is a newer generation. In regards to this generation, one manager also raised the issue of mental health (Interview with PSC manager, June 2018). Many armed responders do not only come from less wealthy homes. They also often come from unstable homes and that instability might continue into adulthood. According to this manager 'half his time is spent counselling' troubled employees. This might be exacerbated by the trauma armed responders sometimes experience on the job. While accompanying workers on patrol some openly share stories of their personal and professional trauma. It typically, would not take long for an armed responder to tell me how often he has been shot at or even hit by a bullet. It also did not take long for many to bring up the, then recent, incident of an armed responder being shot dead in the neighbouring Klerksdorp. During a night shift, an older armed responder started talking about an instance where he was nearly shot dead at close range. I did not do much to probe the discussion. I just let him speak while we were driving. He eventually took me to the exact spot in an outlying suburb, where this took place. He stopped the vehicle in exactly the same spot and orientation in which it was idling when the incident happened and pointed to the spot from where the shots were fired. Luckily they missed. If he had stopped 'a centimetre to the right', he would have been shot. Because of how the bullet ricocheted, a passenger, had there been one next to him, would have been dead, he said. Later that night I brought up an instance that happened a few years prior, where a lady was murdered in a neighbourhood on the edge of the city. Again, the discussion seemed to take him back to that moment. I did not know that he was on the scene that night and that he frantically tried to resuscitate the lady. He mentioned how he 'lost it' that night. Clearly armed responders require many things, that might be lacking at this stage. More money is certainly one. Trauma counselling is another.

Armed responders sometimes have to face difficult confrontations between home or business owners and suspects. In one instance a young man wanted to rob a shop on the Bult. The owners caught him and wanted to kill him. The PSC had to intervene, but this was very stressful, as the shop owners also wanted to take on the PSC for intervening. In another instance the owners of a scrapyard caught the suspects who apparently broke in. It was an extreme and terrifying challenge to prevent the owners from severely harming the suspects. Because the industry is in the service of paying clients, reporting the such behaviour by clients to the SAPS would be difficult and it may not be done.

Armed response can also be an extremely boring job. It involves driving around the city past customer homes and at times stopping somewhere to rest. Shifts are 12 hours and everyone works a series of day shifts and a series of night shifts, typically three of each, separated by a rest day. Some expressed their frustration with the tedious nature of the job. They would often go an entire shift with merely a handful of alarms. It happens reasonably often for the smaller company I accompanied on patrol, that an entire 12-hour shift would go by without a single alarm. Furthermore, 99% or more of call-outs are false alarms. An insect might fly past an alarm sensor. A window might be partially open and the wind might blow a curtain in front of a sensor or a child might be playing with a panic button. This tedious nature of the work adds to the frustration and low morale of people paid low salaries, as they feel they are not doing anything stimulating or worthwhile with their time and their lives (Hafeale, 2018:2).

All of these factors might help explain why there is a lot of staff turnover in the sector (cf. Hafeale, 2018). According to conversations with members of the industry it is quite uncommon for an armed responder to work at one company for more than two years. An employee might leave company X, to work at company Y, which offered more money at the time, only to be lured back to company X a few years later. Other reasons for staff turnover include the fact that it is a 'thankless job', in terms of how clients sometimes treat armed responders. I will return to this issue in a subsequent section.

Diversity within the industry

Not all PSCs are alike. Some are far more militaristic in their approach and how they project their image through promotional materials than others. Those companies would openly display their combat training on social media. One company uses images of a fleet of vehicles, uniformed workers standing to attention and a branded helicopter to show off its 'credentials'. This open display of might links with Diphoorn's (2015) notion of 'bodily capital', whereby the physically imposing nature of armed responders along with the weapons and bulletproof vests they wear give them currency as crime fighters and in the interpersonal politics associated with their work. Such bodily capital is, however, far from uniform in Potchefstroom, even with companies who pose in this manner on Facebook. Some armed responders don't wear bullet proof vests. Others do not wear full uniforms and others choose to not wear regular firearms guns. Instead they use non-lethal pepper guns.

PSCs also overtly frame themselves as 'community organisations'. For example, the one PSC, which especially draws on this narrative, bears the slogan 'from the community for the community'. One of the armed responders working for this company, I accompanied on patrol, told me that they do a lot in the city that they do not charge any money for. The very first callout we went to was to turn a quadriplegic patient, a former doctor, in his chair, to prevent pressure sores from developing. Within minutes of leaving the doctor's house, we drove past a motor vehicle accident. There were no traffic police on the scene yet. The armed responder I accompanied proceeded to park our vehicle, put on a yellow bib and direct traffic until the authorities arrived. As he was about to start directing the traffic a middle-aged white man stopped him and said, in Afrikaans, 'we appreciate all the good work you do in town'.

Managers and owners too are not all alike. Some are more prone to convenient essentialist interpretations of their employees and segments of society, while others are more careful, at least as regards their 'front stage behaviour' of impression management (Goffman, 1959). According to one manager, the backgrounds of armed responders differ significantly between white and black applicants. According to him some white armed responders typically wanted to go to the army. For them private security is the best alternative. He notes that, 'The whites are 80% out of broken homes. These are people who are looking for a type of 'safe haven' and who want to provide a service for someone else' (Author's translation). He also stated that 'Blacks are different. Their culture is different. I can tell you that a Tswana (a local ethnic group) culture does not work well in this industry' and 'for them it is just a job, maybe 10% see it as a calling' (Author's translation). I do not think, at current salary levels, that it is a surprise that this type of work is 'just a job'. Black armed responders often face additional challenges working in this industry in Potchefstroom. White clients often express dissatisfaction when serviced by a black armed responder. Because of this racism, a white armed responder informed me that, the company he works for is losing clients because they appoint black officers. The convenient essentialism I refer to here might therefore be a rationalisation for demographics within the industry and possibly also for the paternalism I will deal with shortly, which in turn sustains the hierarchies between owners, managers and workers.

According to owners and managers interviewed, some, especially younger men, joining the industry might be a bit 'hot-headed'. They are often reprimanded. One of the companies' representative informed me that they quickly act to arrest this type of behaviour. 'You teach them responsibility', he noted (Interview with PSC owner, July 2017). The headquarters of this company is in Klerksdorp. The owners' offices are under a thatched *lapa*. This type of structure is often found in the backyards of a middleclass house. It may or may not have walls, but its defining characteristic is the thatched roof structure, typically used for socialisation over weekends. These offices appear to have once been a rather large house with a lapa at the back. Other extensions were clearly made and neighbouring plots and the structures on those complete the complex that is the company's headquarters. Employees would tell me, that they know they are in trouble when they are summoned to the lapa. But, they also told me that, unless you tested positive for an illegal substance (there are regular tests), which is not tolerated, you are typically fine if you are honest. There seems to be respect for management and how they treat workers. Management is considered strict, but not unreasonable. Baring exceptional circumstances, armed responders are not allowed to speed. They usually have to pay any speeding fines they get. Moreover, vehicles are remotely monitored. Owners and managers often respond to poor driving with a reprimand over the phone or radio. There are therefore consistent attempts by management in PSCs to ensure that workers act in a professional manner. There can be little doubt that these measures relate to the image of the company, but it nevertheless attests to an understanding within the industry of what responsible action entails, at least in part and even when not everyone within an organisation adheres to such principles.

Armed responders often hold orientations towards their work pertaining to matters of responsible conduct. Some armed responders outright refuse to carry a lethal firearm. Unarmed private security in itself is not unprecedented. Typically, the focus is on unarmed guards in different parts of the world, be it resource extraction in African countries (Abrahamsen and Williams, 2009) or guarding in New Zealand (Bradley, 2020) amongst various other contexts. The choice offered to armed responders and the fact that some choose not to carry lethal weapons is, however, less frequently reported. One comparable exception is Fabbri and Klick (2021) who explore and question the efficacy of unarmed patrollers in the USA. In Potchefstroom, armed responders who choose to not carry lethal weapons work for the largest PSC in city. They carry the aforementioned pepper guns. Based on my discussions during patrols their choice is based on one or both of the following reasons. Firstly, for some the responsibility of carrying a lethal devise, and the potential consequences if they made a mistake, is just too much to bear. They simply do not want to take a life, especially not by accident. Others are weary of being outnumbered in a situation where their firearm is taken from them and they are shot with their own firearm or where they lose it and it becomes a tool for criminal activity. Consider the following statement from a recording of a patrol session.

> Actually I hate this bloody thing (the gun). Let's say I get involved in a fight. There are three or four people. They grab me and take my gun from me and shoot me. What can I do? Nothing.
>
> (Author's translation)

Carrying firearms is a disputed matter amongst those working in private security in Potchefstroom. Owners and workers of other PSCs, state that they advertise armed response and as such they are obligated to provide this service. They would rather focus on frequent refresher training. But, according to them carrying a non-lethal devise does not constitute being armed. For example, one armed responder explained to me that he would not want to work the same shift as someone who is not carrying a lethal weapon. His colleague would not be able to provide him with the necessary backup if he really needed help. This company only has two vehicles that patrol Potchefstroom on a given shift. Security companies who do force workers to carry real guns do a lot of training. Chapter 2 of the *Firearms Control Act* of 2003 and The *Private Security Industry Regulation Authority (PSIRA) Act 56 of 2001* provide guidelines in this regard. Some of the key requirements are that both the security worker and the company have to be registered security providers. Each armed responder has to be in possession of a certificate of competence issued by a recognised authority, which may be a private company. At a minimum, to retain firearm competency, practical training is required at least every 12 months. Workers must have a 'stable mental condition' and not be 'inclined to violence', although no professional medical assessment is required in this regard.

There is a huge responsibility associated with carrying a gun and generally those I have dealt with are very aware of this.

As alluded to previously, some PSCs offer additional services well above and beyond security. Thus 'community service' takes on both positive and contestable forms. PSCs have in fact become a *de facto* number for all emergencies, including medical emergencies. One owner noted: 'People don't know what to do in medical emergencies' (Interview with PSC owner, February 2017). They can call the security company. The armed responders are trained to stabilise the patient on the scene and the company in question has a formal agreement with a local ambulance service and hospital. They have radio communication with the ambulances. They get the ambulance to the scene and they have the clients' medical records, including allergies and their doctor's details, which they provide to the paramedics. Other companies also have their own paramedics and/or trained their armed responders in first aid. Then there are incidents they respond to, which strictly and legally speaking are not within their purview. Sometimes, often where clients complained, PSCs have to respond to non-criminal complaints such as disturbing the peace. In these instances, they have no legitimate authority and their approach is a type friendly negotiation.

One PSC as part of their community service function cooperates with a local municipal councillor in providing safe transport home for female students who find themselves in 'trouble'. Here, the sense of 'moral decay' one can glean from Chapters 2 and 4, manifested during an interview (Interview February, 2017). The owner stated that 'These girls do shocking things. And then it is girls who come from "normal" homes' (Author's translation). In these instances, seemingly, religion-based counselling is given. Some PSCs do their own investigations and draw up profiles of prominent criminals, and they arrest people technically for offences they have no jurisdiction over, such people carrying drugs. The fact that such arrests continue suggests that the SAPS do sometimes process these cases. In particular arresting people for offenses that do not fall within Schedule I of the *Criminal Procedure Act* (South Africa, 1977) is not allowed. Schedule I offences mostly include immediate threats to bodily integrity, such as murder, attempted murder, sexual assault, assault with the intent to do grievous bodily harm and armed robbery, but also breaking and entering and malicious damage to property. Arrest may take place if there is a 'reasonable suspicion' that a person committed a Schedule I offense, which of course creates openings for PSCs to overstep, should they choose to. As mentioned in Chapter 2 and reported by Van Riet (2020:90) PSCs are often summoned by clients to police vagrancy and harass and remove homeless people from middleclass suburbs based on dubious accounts of 'suspicious' behaviour. Schedule I also provides guidelines on the use of force, which may only be sufficient to restrain the suspect and for the purpose of handing the suspect over to the SAPS. There have been court cases where PSCs were accused of acting unlawfully, while the suspect pursued might go free.

Along with the 'community service' cited earlier another floating signifier worthy of scrutiny is the often-used term of a 'force multiplier' that PSCs often describe themselves as. Abrahamsen and Williams (2005:431) as far back as 2005 draw on this concept in a constructive way, as a tool towards providing security for the public more broadly defined. In the JB Marks Municipality, this is precisely not what is happening. Only those who can afford monthly subscription or at least those who live in areas frequently patrolled reap the benefits on private security. The notion of a force multiplier may also be problematic where PSCs move outside the remit of its rights and obligations. As PSCs essentially have the same powers of arrest as an ordinary citizen these parameters are relatively narrow. Yet, overstepping the bounds of their authority does happen, as mentioned earlier. In another instance, a PSC alerted the police to a particular drug house, reportedly repeatedly so. The SAPS did not do enough in their eyes. Then one night members of the PSC went into the house, broke the television set and took the drugs.

As alluded earlier, patriarchy is a significant feature in the routine operations of at least two of the larger PSCs in Potchefstroom. The one company employs the car guards who work at two of the shopping centres in the city. Its manager noted in a rather condescending and again, essentialist, manner that:

> These are the type of guys who really live from day to day. They don't want a monthly salary, because they cannot manage it. He does not want to work permanently.

He also stated that if one of these car guards stands out and demonstrates 'an ability for greater responsibility', they will make him an armed responder and 'if he leaves us, then we want him to get something better'. Another company offers short-term interest free loans to workers in distress. They do this instead of paying their workers higher salaries. Men own these companies. They do the managerial work. Women are receptionists, secretaries and work in control rooms (along with men). This is interesting. In the SAPS women fulfil all roles. What they lack in relative physical strength, they are meant to make up in training. This was even the case when some of these PSC owners and managers were in the police decades ago. These interventions towards upward social mobility for those who 'prove themselves', side-by-side with to the essentialist and at times racist discourses might be interpreted to similar utterance of care towards 'troubled workers' discussed previously. It is perhaps best understood as a form of gatekeeping that justifies existing racial, class and gender-based hierarchies within the industry.

Relationships with the public

Private security companies are much hated amongst a significant proportion of the black population. These companies are often viewed as serving white interests. Some armed responders told me that there are parts of the city where they would prefer not to drive through, with a company vehicle, because of the response that it might illicit. One armed responder explained the difference between students when he is called out to complaints of disturbing the peace. He noted that white students would say 'sorry, I know I am drunk, I did not realise I was this loud', while black students would call him a racist. This comparison reflects pacifism, associated with a history of Calvinist socialisation, in the wake of apparent authority by many white Afrikaans people. It is also clear that many residents know their rights and are unwilling to tolerate questionable authority, especially from a set of organisations historically associated with protecting elite interests.

Linked to the matter of race, is how clients treat armed responders. As mentioned, many clients do not like being assisted by black armed responders, to the extent that they will move their business to another company if they are serviced by a black officer. This places pressure on PSCs to work against the social transformation that is legally mandated through the *Employment Equity Act* (South Africa, 1998). Consequently, white armed responders are the overwhelming majority of such workers in the city. In addition, they are often asked to perform frontier governance and thus preserve a problematic, segregated, status quo through requests to respond to 'suspicious' activities, as mentioned earlier.

Clients have demands and if these are not met, they can always move to another company. A competitive rate is simply one of these demands. In Potchefstroom a standard armed response subscription might cost a household around ZAR 300 per month. Sometimes there are specials and a free standard alarm installation is included when you sign a contract for a period of time. Because of competition, PSCs don't always have much room to manoeuvre on subscription charges and salaries, unless they add value in other ways. This is where synergies with other infrastructures such as the CID and the roll out *en mass* of CCTV cameras in the city can be a major factor especially in favour the largest PSC. Another competitive advantage this company has is through the financial services company that forms part of the larger group to which it belongs. An armed response subscription allows the insurance brokerage business to charge lower premiums to clients of this particular PSC. Other demands can at times be quite unreasonable. Clients would be rude to armed responders of all races who in their eyes arrived late. Most alarmingly, they would expect PSCs to do what the SAPS are unwilling to do, on account of constitutional rights such as freedom of movement (see Chapter 2). There are of course exceptions.

We might now consider what it is clients pay for. As armed responders, as a rule, are not allowed to speed, response time dictates that PSCs cannot easily prevent certain forms of crime, such as housebreaking and house robberies. They do however offer an incentive for criminals to leave the area quickly. This might at times limit the potential for harm to life and limb.

The relationship with the SAPS

Depending on who you speak to, the relationship between PSCs and the SAPS in Potchefstroom has, historically, largely varied. This can be based on the individuals in charge, in other words, who the station commander is (Interview with

PSC owner, February 2017). Others note that although there is a large demand for armed response, because 'you cannot rely on the police anymore', there are at times very good relationships with the SAPS (Interview with PSC owner, September, 2019). As mentioned in previous chapters, the new station commander's role in this regard appears to be significant. He has constructively engaged residents and security actors in his precinct while being relatively firm on all dimensions of the rule of law. For example, in a CPF meeting I attended he praised residents for initiatives whereby they are 'taking back their neighbourhood', by establishing a presence in public spaces, while being very firm on the limits of acceptable actions, that may be taken in this regard. There is a weekly meeting between the SAPS and PSCs in the city every Thursday morning, where information is shared. As with the presence of unarmed private security actors, this too is not unprecedented. Abrahamsen and Williams (2007:246–247), for example, noted collaboration between the SAPS and private security in Cape Town. Not only are public–private partnership a common phenomenon in security provision post-apartheid, through the likes of CIDs (see Chapter 9), but the case of Potchefstroom suggests, as Govender (2017:2) argues, that more informal collaboration also takes place and it might ebb and flow, as relationships are at times strained. Because PSCs have many more vehicles on the road patrolling at any given point in time, they often have a lot to offer the SAPS, especially in terms of information. For reasons explained in the following two chapters, a situation where PSCs dominate security provision in a given area can be problematic. In short, frontier governance, becomes more pervasive, as PSCs primarily cater to paying clients, who exist in vast amounts on account of structural causes of crime.

There appears to be an understood approval given to the PSCs to take some of the responsibility for crime prevention off of the SAPS's hands, through patrols and doing visible policing. Govender (2017:3) however notes that this is a grey area, as the legal powers of PSCs in this regard is ill-defined. Again the notion of a 'force multiplier' comes to mind. There does sometimes appear to be a willingness on the part of the official force (the SAPS), for others to be their 'multiplier'. One PSC owner told me that they will never compete with the larger companies in the country like Fidelity, which in recent years took over another giant ADT. His objective, instead, is to become the 'largest small security company' in the country. He would also like for his company's number to replace the police's 10111 number (Interview with PSC owner, July 2017). The 10111 number can be compared to the 911 number in the USA. It is a general emergency number. This company reportedly has the largest radio network in the southern hemisphere. It stretches some 500 km². They monitor eleven different emergency services' radios. This includes, the fire brigade, ambulances, tow trucks and other services. Private security companies also sometimes have SAPS radios in their vehicles, in cases where there is more open collaboration between them and the SAPS.

Importantly, armed responders in Potchefstroom tend to understand their role as one with less authority than the SAPS. They might be first of on a scene, but they do not do forensics or analysis of any kind. They secure the scene to hand it over to the police. In this regard they seemingly accept the 'junior part-ner' role (Diphoorn and Berg, 2014:426) in their relationship with the SAPS who have more legal powers. When they make an arrest they hand over the scene and the suspects to the SAPS. In many instances the crime scene has to be preserved. Subsequent to a farm murder just outside Potchefstroom, an armed responder explained to me that he did not want the scene to be contaminated, by himself or anyone else. 'I only cordoned off the victim's body with danger tape and basically just looked after the body until the police arrived' (Author's translation).

Generally, PSCs also appear cognisant of the legal limitations of the services they can provide. Owners and armed responders would state outright that they have the same powers of arrest as ordinary citizens. The exceptions stated in the previous section, where PSCs completely overstep their mandate, notwithstand-ing, in general, PSCs operate within the prescribed parameters. Sometimes they do push these limits, by gathering intelligence on their own and through informal networks. They might access to cell phone data obtained through an official war-rant, car number plate recognition and facial recognition. Much of this access depends on the quality of the relationship with the SAPS more formally, which ebbs and flows, and informal back channels to the SAPS.

Given the limited resources of the SAPS and the relatively well-resourced nature of PSCs this crossing over of PSCs' work into the SAPS' mandate is under-standable, although it is problematic for the legal reasons already cited. Security companies do fulfil an important role when the SAPS are stretched too thin. In the case of Potchefstroom, if there are two fatal accidents at the same time on roads into and out of the city, the SAPS may, in theory, not have any vehicles to dispatch elsewhere. According to various sources, there are at any given point in time only two to three vehicles available for call-outs. Moreover, when someone dies in a vehicle accident, the legal procedures assigned to the SAPS become far more involved. As such, these two hypothetical accidents would occupy the offic-ers from both vehicles for numerous hours.

Security companies echo many CPF members and citizen concerns in terms of the SAPS. There is a sense that although there might be a shortage of resources, it can still be deployed more productively. Having stated this, there appears to be a lack of understanding or rather appreciation for the 'triage' that the SAPS are forced to utilise to allocate resources. For example, they respond much faster to robberies and crimes against persons than ordinary burglaries. It is therefore, understandable that the SAPS might take long to respond to a burglary, when they are on the scene of a murder or motor vehicle accident where someone died. Consequently, in a case of theft from a client, say a shop, the PSCs often have to dedicate resources for up to five hours to hold the suspect before the SAPS arrive and the case can be handed over to them. This is an understandable frustration on the part of the PSCs.

An obvious question to ask now concerns implications of the aforementioned understood approval given to at least some PSCs for aspects of crime prevention.

Krahman (2008) notes that there is a change in the nature of security provision when it is viewed as a commodity and not a public good. These companies have historically had a different mandate, mostly of response to crime. They are led by the needs and, as mentioned, often unreasonable demands of clients who comprise only a numerical minority of the population. What happens when these demands penetrate the perspective of private security workers in a context where they have an understood approval for patrol and prevention? Such a concern seems to apply mostly to the larger companies. As mentioned, the smaller companies stick to a simple formula of alarm systems and armed response. Based on analyses from previous chapters we might ask whether it is not because of the growing gap in terms of trust between the SAPS and the community (elaborated in Chapters 2 and 4), that PSCs are drawn into a vacuum of sorts? That might indeed be partially true, but this type of concern can only be legitimate in as much as the demands of the public are reasonable and legal. Moreover, private security in an unequal context such as South Africa implies geographic, class and race-based disparities in security provision.

There seems to be, in the aggregate, an understanding within the industry of crime that can be reconciled with the explanation provided in Chapter 1. There are however differences amongst members of the industry. One owner, for example, lamented that there is a 'non-existent legal system'. 'There is no fear' amongst criminals. There is a 'bad government'. Now there is a 'communal feeling of law-lessness'. 'It's a sick society.' 'It's not always the poor person who does crime.' There is a 'lack of social cohesion' (Interview, PSC owner, February 2017). Other owners and managers tend to acknowledge, as more significant, issues around a lack of legal options for earning and income. One owner even emphasised his astonishment at how much risk more organised groups or robbers take for very little reward (Interview, PSC owner July 2017). The armed responders I dealt with are near universally understanding of poverty and inequality as drivers of crime. As the aforementioned comments seem to show understandings of crime as linked to poverty, inequality and a lack of social cohesion do manifest significantly within the private security industry in Potchefstroom. Therefore, security com-panies, as significant security infrastructures, might be engaged productively by using this relatively common language, while advocating for the rights enshrined the Constitution (South Africa, 1996). To complicate this discussion further, and to point out how individuals are complicated and not uniform actors, I can note that the same more conservative owner referenced earlier noted that 'we will not do arrests with violence' and 'we will not act outside the law' as 'this will simply fuel hatred'. And that 'zero tolerance has its place, but it should be located in the criminal justice system, not private security'. Comments such as this might also be a front stage performance, although the fact that it was interspersed with less sympathetic comments suggests otherwise. What the evidence however suggests is, at a minimum, that PSC managers and owners of the largest and most signifi-cant PSCs in Potchefstroom have a fair idea of what are appropriate forms of front stage behaviour. This realisation gives diverse interest groups something to work with when critically engaging these organisations.

To conclude this section, it should be stated that various authors writing on private security have rightly noted how the Weberian ideal-type state is just that, an ideal-type. The state has never truly held a monopoly on legitimate force (Abrahamsen and Leander, 2016:1; Shearing and Stenning, 2016:142). Some have suggested more horizontal notions of security provision and accountability, such as 'hybridity' and 'simultaneous authority' as more in line with contemporary realities in security provision (Berg and Howel, 2017; Shearing and Wood, 2003). While this is true we need to consider, within the context of this book's analysis, *what type of hybridity* is desirable to maximise the accountability of private security providers. Security provision should not be the preserve of the minority at the cost of the majority and as such there are potentially limits to a horizontal understanding of security provision and accountability. By this comment I do not mean to impute upon the cited authors a normative argument in favour of such horizontality. They too are grappling with an interpreted empirical reality. There is however potential in the notion of an understood mandate to conduct aspects of crime prevention, that can be leveraged in favour of policing based on democratic sensibilities. Following Linz and Stepan's (1996:10) conceptualisation of the consolidation of democracy, I conceive of policing by democratic principles as policing that will likely enhance or at least not harm social cohesion, through perceived double standards. This implies that hybridity and border governance, as defined in Chapter 1, are not compatible. Private security provision in Potchefstroom is by no means perfect. However, even though the perennial danger remains of private security serving as a vehicle of the powerful, there are nuances and variations within the industry and cited in this chapter. These nuances suggest at least a semblance of hope for rearticulating the social towards less divisive ends.

Conclusion

As per the Laclauian approach that informs this book, a few significant sutured categories and floating signifiers have emerged from this chapter. These signifiers are not all equally problematic, but they all have the potential to justify dubious activities by PSCs. Such dubious activities might not even be intentional. The notion of 'community' reads as a marketing ploy. We may legitimately ask, 'who is this community' and clearly it is a community of clients, which is mostly limited geographically. In the neighbouring Ikageng, far fewer residents are able to afford armed response subscriptions. One way of rearticulating 'community' might be to have a vehicle, even a vehicle shared between companies and the SAPS, that is dedicated especially to the outlying newer extensions of Ikageng. Competition between the SAPS and PSCs and between PSCs might be cited as reasons why this cannot happen. So too may a lack of funds. However, by sharing this responsibility, the expense might be smaller and importantly the image of PSCs as divisive organisations within society might be changed. Furthermore, if the SAPS and one of the PSCs in Potchefstroom can reach an agreement on a CID in Potchefstroom, then why can they not do the same where people are most vulnerable, especially to violent crime. To help such a project along might require one or more corporate social responsibility partners, just like the NWU has become a major source of funding for the CID (see Chapter 9).

As a second floating signifier, this chapter has shown how race is still a major problem in Potchefstroom. Some PSC workers and their clients display racism. Sometimes this racism is situational and sometimes there might be behaviour inadvertently reinforcing racist orders and stereotypes. It is important that, especially those who fall in the latter group be engaged constructively, to form a larger critical mass in opposition to more overt racist practices. A major part of such rearticulation would be a means by which to contend with the often unreasonable and illegal demands of clients. Part of the solution might be to force interaction across racial divides, by for example forcing PSCs to transform the demographic composition of their workforce. This may also help reduce the partially race-based patriarchy that is found in some parts of the industry. The *Employment Equity Act* (South Africa, 1998) already offers guidelines on workforce transformation. Again the Lefebvrian (cf. 1990) tension between spaces of representation as a conservative attachment to narratives of crime and representations of space, which ought to be more inclusive are apparent. The continued enforcement of exclusive spaces of representation has a self-perpetuating quality that must be arrested. For this to happen, reform within PSCs could help, but there also needs to be an improvement in the crime prevention capabilities of the SAPS. This will reduce crime and help to isolate and call-out racist tendencies that have endured within society.

As a third floating signifier, the notion of a force multiplier could be a very useful way of looking at the role of PSCs. They do have a different mandate, but they have a bigger footprint on the streets. They can be a significant ally to the SAPS as a 'junior partner'. In another sense, it would appear, that the notion of force multiplier has at times developed into something more akin to a force *replacer*. The problem seems to be both in terms of the capacity of, and trust in, the SAPS, that leaves a gap in crime prevention and response that needs to be filled. On the other hand, this matter also relates to some clients' unreasonable demands. Hence, there appears to be much norm confusion in society, where one form of injustice can seemingly only be addressed through expedient and sometimes unlawful behaviour. If there is something wrong with a black person walking in a particular part of the city, then clearly there is something wrong with how the society in question (dys)functions.

There is however hope in the diversity within this industry. Firstly, a basic understanding of the socio-economic roots of much crime creates at least a partially common vernacular through which more democratic policing strategies may be discussed, where they do not yet exist. The backgrounds of armed responders and their socio-economic status means that this insight is not lost on them. Secondly, the notion of hot-headedness and macho personas, as a characteristic of PSC operatives is often questionable. Unlike studies in larger urban areas, armed responders in Potchefstroom seem to exude less 'bodily capital' (cf. Diphoorn, 2015). There is certainly no major sense of urgency in responding to call-outs.

Some are not fully uniformed all the time and, as noted, many armed responders choose not to carry lethal weapons.

So what do we make of Lisego the security worker introduced at the very beginning of this chapter? I would argue that he is not typical. There are people within Potchefstroom's private security industry who have at times acted like him and advocated for similar actions. These instances are however limited. No person is just one thing. All the workers within private security are complexly composed through their collective frames of reference and the power relationships that inform their daily realities. By emphasising the sutured categories and floating signifiers cited earlier, including the very notion of an 'armed responder', activities like those committed by Lisego could be minimised. Simultaneously, higher level political-economic changes and greater capacitation of the SAPS and the courts in dealing with crime could reduce the need some feel for acting in this way. This type of enhanced functionality of the broader economy, social services and the criminal justice system might more frequently dissuade PSCs from overstepping their mandate. Moreover, greater demographic transition of the industry in Potchefstroom, might help curtail the racist demands of some clients. Such change might have to be enforced at a higher level, as the transformation of one company might simply mean that clients move to another.

Note

1   These currency conversions were done on 24 March 2020. We should bear in mind that this was well into the Covid-19 crisis. Many investors had already retreated from emerging markets and the value of the South African Rand had fallen to a level very close to its all-time low against the US Dollar.

References

Abrahamsen, R. and Leander, A. 2016. Introduction, in R. Abrahamsen and A. Leander (eds.) *The Routledge handbook of private security studies*. London: Routledge.

Abrahamsen, R. and Williams, M.C. 2005. The politics of private security in Kenya. *Review of African Political Economy*, 32(104/105), pp. 425–431.

Abrahamsen, R. and Williams, M.C. 2007. Securing the city: Private security companies and non-state authority in global governance. *International Relations*, 21(2), pp. 237–253.

Abrahamsen, R. and Williams, M.C. 2009. Security beyond the state: Global security assemblages in international politics. *International Political Sociology*, 3, pp. 1–17.

Abrahamsen, R. and Williams, M.C. 2010. *Security beyond the state: Private security in international politics*. Cambridge: Cambridge University Press.

Berg, J. and Howell, S. 2017. The private security complex and its regulation in Africa: Select examples from the continent. *International Journal of Comparative and Applied Criminal Justice*, 41(4), pp. 273–286.

Bradley, T. 2020. 'Safe' and 'suitably qualified': Professionalising private security through mandatory training: A New Zealand case study. *International Journal of Comparative and Applied Criminal Justice*. http://doi.org/10.1080/01924036.2020.1719528

Chrisholm, A. 2016. Postcoloniality and race in global private security markets, in R. Abrahamsen and A. Leander (eds.) *The Routledge handbook of private security studies*. London: Routledge. pp. 177–186.

Diphoorn, T. 2015. 'It's all about the body': The bodily capital of armed response officers in South Africa. *Medical Anthropology*, 34(4), pp. 336–352.

Diphoorn, T. and Berg, J. 2014. Typologies of partnership policing: Case studies from urban South Africa. *Policing and Society*, 24(4), pp. 425–442.

Dzhingarov, B. 2017. Cele's private security industry regulation bill meets opposition. *BizCommunity*. Online: www.bizcommunity.com/Article/196/181/162964.html. Date of access: 21 March 2021.

Fabbri, M. and Klick, J. 2021. The ineffectiveness of 'observe and report' patrols on crime. *International Review of Law and Economics*, 65. https://doi.org/10.1016/j.irle.2020.105972

Foucault, M. 2003. *Society must be defended: Lectures at the Collège de France, 1975–1976*. New York: Picador.

Goffman, E. 1959. *Presentation of self in everyday life*. Garden City, NY: Double Day Anchor Books.

Govender, D. 2017. The role of private security in crime prevention. *Servamus*. November, pp. 30–32.

Hafeale, C. 2018. Assessing the levels of flourishing in a private security concern. Mini-dissertation for the degree Master of Business Administration. North-West University. Potchefstroom.

Krahman, E. 2008. Security: Collective good or commodity? *European Journal of International Relations*, 14(3), pp. 379–404.

Laclau, E. 1990. *New reflections on the revolution of our time*. London: Verso.

Laclau, E. and Mouffe, C. 2014[1985]. *Hegemony and socialist strategy*. London: Verso.

Linz, J.J. and Stepan, A.S. 1996. *Problems in democratic transition and consolidation: Southern Europe, South America and post-communist Europe*. Baltimore: Johns Hopkins University Press.

Shearing, C. and Stenning, P. 2016. The privatization of security: Implications for democracy, in R. Abrahamsen and A. Leander (eds.) *The Routledge handbook of private security studies*. London: Routledge.

Shearing, C. and Wood, J. 2003. Nodal governance, democracy and the new 'denizens'. *Journal of Law and Society*, 30(3), pp. 400–419.

Smallhorne, M. 2017. You're not above the law, Mr Armed Response in *Fin24*. Online: https://m.fin24.com/Opinion/youre-not-above-the-law-mr-armed-response-20170619. Date of access: 26 March 2020.

South Africa. 2003. *Firearms control amendment act 43 of 2003*. Pretoria: Government Printers.

South Africa. 2001. *Private security industry regulation authority (PSIRA) act 56 of 2001*. Pretoria: Government Printers.

South Africa. 1998. *Employment equity, act 55 of 1998*. Pretoria: Government Printers.

South Africa. 1996. *The constitution of the Republic of South Africa. Act 108 of 1996*. Pretoria: Government Printer.

South Africa. 1977. *The criminal procedure act, act 51 of 1977*. Pretoria: Government Printer.

Van Riet, G. 2020. Intermediating between conflict and security: Private security companies as infrastructures of security in post-apartheid South Africa. *Politikon: The South African Journal of Political Studies*, 47(1), pp. 81–98.

Van Tillburg, L. 2019. Clifton beach incident highlights problems with private security in SA in BizNews. Online: www.biznews.com/leadership/2019/01/16/clifton-beach-private-security-sa. Date of access: 26 March 2020.

Interviews cited

Police officer. Interviewed by Gideon van Riet. 9 August. 2018.

Private security company manager. 2018. Interviewed by Gideon van Riet. 7 June. 2018. Potchefstroom.

Private security company owner. 2019. Interviewed by Gideon van Riet. 19 September 2019. Potchefstroom.

Private security company owner. 2018. Interviewed by Gideon van Riet. 20 February 2017. Potchefstroom.

Private security company owner. 2017. Interviewed by Gideon van Riet. 7 July 2017. Klerksdorp.

8 Surveillance in the suburbs

The pilot project in Oewersig[1]

Introduction

This chapter deals with the work of the Oewersig Residents' association. It is an organisation officially set-up to deal with crime, neatness and the provision of a 'nice' or 'good' living environment for residents of this up-market neighbourhood. Residents' associations are not unique to Potchefstroom, although it is the first such instance in the JB Marks Municipality. To some extent, Oewersig mirrors some of the findings in other settings such as Johannesburg, where these associations have been set up to deal with 'crime and grime' broadly conceived (Clarno, 2013:1198). Relative exclusion and what I have conceived of as techniques of frontier governance follow from the fundamental changes that came with democratisation. Similar to the simultaneous logics of semigration and recolonising public spaces, described in Chapter 2, elites responded to democratisation by 'jumping scales' in attempts to maintain terrains of dominance. Residents' associations, gated communities and CIDs are all 'scalar fixes' that shores up parts of the city as a base of operations, which might include further expansion, even if not by the exact same members of elite groups Clarno (cf. 2013:1198). Scalar fixes, such as the Oewersig Residents' Association represent attempts at sustaining spaces of representation in Lefebvrian terms, that caters more to a specific segment of the social order compared to others (Lefebvre, 1991:1–67). I will, however, argue that although this has been the effect, it might not entirely have been the intentions of all involved. The situation is more complicated and as such there remains room for meaningful engagement with at least some residents of this neighbourhood.

Oewersig is a relatively easy neighbourhood to police through live CCTV camera monitoring. There are only two street entrances into the neighbourhood laid out in a horse shoe formation, with side streets. On the one end of the ring road is the Mooi River. The area around the river is a public park and frequented by students and picnickers. Oewersig is an important case study. Other neighbourhoods and the CID discussed in the following chapter have started to implement similar security measures. In addition, Oewersig represents a significant political dynamic between spaces of representation envisaged by residents and representations of space, sans apartheid, as envisaged by the city council. It is clear that no single group got their way entirely and at least some representations were upheld in the name of democratic values.

DOI: 10.4324/9781003028185-12

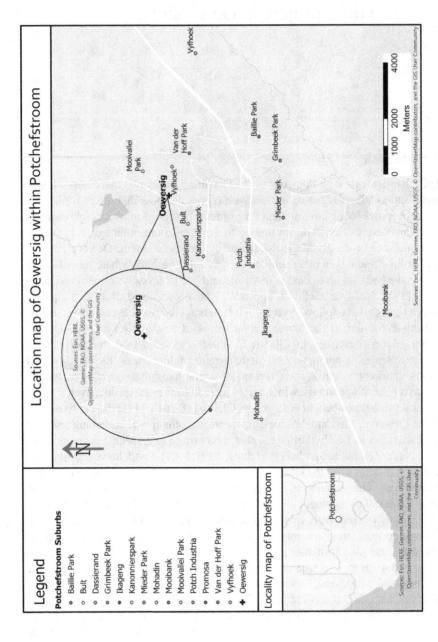

Map 8.1 Location map of Oewersig

Again, as has been argued throughout this book, hope might be found in the lack of homogeneity and absolute agreement within the group of residents. For example, signifiers such as 'neglect of white neighbourhoods' by the city council should be challenged, while other residents display a more open mindset. That of course does not mean, that this type of security set-up is desirable. Residents, most of whom are house-owners, need to be given more options that are less intrusive to the general non-resident public, yet reassuring for efficacy in crime reduction. The means by which to establish such alternatives, sadly, largely lies at different scales or levels of decision-making. The chapter affirms the argument introduced previously, that there is a potentially toxic gap between state and society that in the case of crime and security is filled with various potentially dangerous and problematic security infrastructures, discourses and practises.

The residents' association has taken responsibility for certain municipal services on account of 'poor (municipal) service delivery'. For the aforementioned reasons and the fact that the police are held in poor regard throughout the city and South African, there is a real danger that most of Potchefstroom will eventually be subject to similar surveillance techniques. Indeed, these technologies of surveillance are steadily expanding towards the South. Preliminary evidence suggests that cameras in the street have already moved, as opposed to stopped, crime. For as long as lower middleclass and poorer households do not have the means by which to purchase similar security they might increasingly suffer crime. To this we may add the obvious ethical dilemma of being consistently captured and viewed on camera by an anonymous individual who works for those with means. This chapter is largely based on interviews with residents as well as weekly 'incident reports' compiled by the PSC in question, which the association shared with me.

The remainder of this chapter is structured as follows. The next section provides a systematic description of the origins of the residents' association. It then scrutinises the potentially problematic discourses associated with the formation of the association and its activities. These relate to political economy and notions of neglect by the state. The chapter then considers the association and its activities as a set of security infrastructures that are intertwined with the politics of space in a Lefebvrian manner. In other words, there is a dialectical politics between representations of space and spaces of representation, that implies both conflict and cooperation on particular matters. Again, these instances of negotiation imply a promising metaphysical or cognitive space for cooperation, in this case between different elites, but against the backdrop of a much more stratified social order. The chapter therefore affirms that hegemony is never a completed power structure, nor is it internally coherent. It is indeed a space where much promise for productive change lies.

Establishing a residents' association

Every house in the neighbourhood constitutes a single member, which makes a relatively small contribution every month. This started at R100 per member in 2014 and moved up to R200 (about USD14 as of 20 June 2019) per month in

2018. Pensioners get a 10% discount. Members have also been able to negotiate discounted insurance on their property as the PSC that monitors the cameras are part of a larger consortium that also includes an insurance brokerage firm. They deliberately started small. It was argued that to get maximum buy in from the neighbourhood, you cannot start with a very large amount.

The association is a non-profit organisation with a constitution and office bearers with portfolios. These portfolios include communications, security and contact with the SAPS; administration; finance; and maintenance amongst others. The finances of the organisation are audited. There is a steering committee that meets monthly. There is also a larger annual meeting where activities for the entire year are reported to the members more broadly. Over 90% of households in the Oewersig are members of the association. There used to be issues with those who rent, who would not pay. At the time of writing there were merely one of two households that did not pay the voluntary fee. The association's approach to these non-payers, reportedly, is not to pressurise them. Instead they continue to send out information and newsletters on what is being done by the residents' association.

Interviewees do not agreement completely on exactly how the residents' association came about. It does appear that there was a gradual process following speculative initiatives around neighbourhood *braais* (as barbeques are known in South Africa). These combined with a confluence of local needs, the emergence of appropriate technologies and a particular triggering event that raised the matter of crime as a priority, shaped much of the association's focus. One participant who was on the municipal ward committee during the initial discussions, stated that he organised a meeting specifically for Oewersig at one point (Oewersig resident 2 Interview, 2018). The SAPS also attended and noted that the neighbourhood actually had relatively low levels of crime. In the beginning initiators found it difficult to develop the association. One participant noted that to get buy-in – you first need to have something (Oewersig resident 4 Interview, 2018). The process from inception to the point where virtually the entire neighbourhood is observable remotely, took about four to five years. These cameras do not point into peoples' yards, but they do cover all of the streets in the neighbourhood.

It would appear, based on some accounts, that a shooting incident in one street was the proverbial final straw that led to increased concerns with security. Housebreakers climbed over a wall into a yard, saw the owner's large dogs and climbed back again. They saw the police which happened to be in the area. Presumably thinking the police were there for them, the housebreakers started firing. This incident apparently made residents realise that they 'had to do something' (Oewersig resident 4 interview, 2018). There have been many other smaller criminal events prior to this one. This was not even the first instance of intruders climbing over the same wall. In previous years this owner saw footprints in the dew on her lawn one morning. However, the shooting incident mentioned earlier was the worst and a type of catalyst for action. This action could take on many potential forms.

Residents entertained many ideas. The initial idea was to put up boom-gates in order to become a gated community and a special rates area. The latter involves an additional levy paid by residents of the designated area to the municipality every month. The additional funds raised then goes towards upgrading the area in question (McCain, 2018). The municipality was opposed to this option in principle and there were no bylaws in place by which to approve it. Then the association considered closing off one of the exit points only and only at night. A PSC would be given the responsibility to lock and unlock this gate. They also considered adding a guard. Both the boom gate and the guard would be very expensive. In 2016 a guard would have cost R12,000 per month, while the 24-hour monitoring of cameras by the PSC costs R2500 per month.[2] This relatively low rate for the live monitoring of cameras is because the PSC that eventually got the contract makes up the shortfall through individual armed response subscriptions in the neighbourhood. Moreover, as explained in the previous chapter, guards can be problematic. They are not paid much, as they only receive a relatively small proportion of the fee charged to the client. They can therefore be easily bribable. In the proposed setup a guard would only cover the one entrance to the neighbourhood, leaving the rest of the neighbourhood exposed. With these limitations and considering the fact that access could not legally be refused to anyone, the association rejected all of these options. They then considered a city improvement district. Again there was a lack of applicable municipal bylaws and again the municipality appeared rather reluctant to consider this idea.[3] One participant stated 'We could not even get them (the municipality) to organise a damn meeting' (Oewersig resident Interview 5, 2018). Based on my interview data, gated community status is still an objective for some, but the association felt that in the shorter term something needed to be done. Moreover, closing off the neighbourhood completely remains highly unlikely as there is, as stated, a public park next to the river. The initial stop-gap solution was fake cameras.

The fake cameras initially erected were however obviously so. Real cameras then became an option which the association pursued, upon advice of the PSC which eventually got the contract. According to various interviewees, they chose the PSC which they 'considered most professional' (Oewersig resident 1 Interview, 2018). Also, as an added advantage the PSC in question is the largest in terms of client base in Potchefstroom by some distance. They are therefore considered more likely to have intelligence on suspicious vehicles and people in the city. The association then started with two real cameras, one at each entrance. These did not render very detailed images. Furthermore, there are many trees in the area which prohibited a proper line of sight for the camera signal to the PSC. Moreover, the speed at which data could be transmitted was problematic and streaming normal internet data would have been far too expensive. Therefore, initially the cameras merely recorded footage. This evidently could not do much for crime prevention. The opportunity to install cameras that could be remotely monitored 24 hours per day came when a local internet service provider (ISP) started to roll out high-speed fibre lines. The particular ISP gave the neighbourhood a preferential rate, which in turn effectively blocked out other ISPs. Cameras were installed in phases, as more funds became available. Some members provided loans to the association for infrastructure projects. The association also put up lights at both entrances. Eventually, with the advent of fibre, high-quality cameras followed with live monitoring by the PSC. There were negotiations with residents who lived next to the cameras for the supply of power. Typically, this would imply, initially, a contribution to the electricity bill and later a separate prepaid meter housed in the yard and funded by the association.

The PSC frequently shares 'incident reports' with the association's management. I am in possession of various such brief reports. These include narratives of potentially suspect behaviour unfolding, the details of the response by the PSC and further details of actions taken. As the number of these reports increased, residents saw the potential value of such live monitoring and they started to give additional donations or loans to the association. This is because 'with limited cameras you can only follow a suspect so far' (Oewersig resident 4 Interview, 2018; author's translation). People were often arrested without residents even knowing about it until the incident reports were distributed. Gradually more cameras were erected in strategic locations according to priority. The neighbourhood is now almost completely covered, with very few, if any, 'blind spots'. One homeowner even installed a camera on his garage wall at his own cost and had it linked to the network.

A WhatsApp group was formed, which management has handled discretely, to prevent unnecessary panic. As a result, only important information about significant threats or evidence of the efficacy of the cameras are shared with the whole neighbourhood. The PSC is part of the WhatsApp group. It might happen that a resident raises a concern and then the PSC would respond by saying 'don't worry, we know that car. They live there' or 'We know this person. He is the gardener at a certain house'. There is no neighbourhood watch. The cameras have made such an initiative redundant with no added danger to residents doing patrols in the streets. They did initially consider a neighbourhood watch, but decided early on that they were not equipped for this. They would not expose themselves to criminals. One respondent explained that they are not properly trained or professional crime fighters. Therefore, the risk for residents to get involved in a neighbourhood watch is simply too much (Oewersig resident 5 Interview, 2018).

The association's work has expanded beyond purely security concerns, to the extent that participants mention the function of the association as a combination of security, neatness and upkeep. Interviewees associate this expansion of functions with an apparent lack of adequate services from the municipality. Residents appear to be ambivalent about this. Some seem torn between a sense of understanding that a neighbourhood such as Oewersig cannot be a priority regarding the immediate needs of society and responsibilities of the city council. On the other hand, they feel that some very basic services are neglected (cf. Oewersig resident 5 Interview, 2018). Others are more critical, even stating that 'the state don't care about white neighbourhoods anymore' (Oewersig resident 3 Interview, 2018).

One of the very early initiatives was to provide plastic 'wheelie bins' to each household for refuse removal. The association also created a recycling system. Every Thursday paper, cardboard and glass are dispensed with. The association used to pay a company to collect these materials, but the amount of recyclable waste eventually reached a critical mass, making it profitable for the recycling company to drive to Oewersig and collects these materials. The association also added educational plaques that provide information on animals, birds, insects and plants found in the area. They remove branches that block the river. They have placed six rubbish bins next to the river. However, according to participants the municipality does not cut the grass anymore and those who picnic litter a lot (Oewersig resident 5 Interview, 2018). They have quarterly clean-ups where residents walk with refuse bags and pick up litter. They regularly rent a tractor and pay for the diesel to cut the grass on the river bank. The association contributed to the erection of a fence between Oewersig and the neighbouring school's sport fields. There still is a bridge over the river between the two areas. The incident reports however, show that many potential criminals still enter the neighbourhood by crossing the river. Other initiatives include buying manhole covers that have been damaged and even intervening when there are complaints about restoring the peace.

The association also see to it that lines and speedbumps in the streets are painted. They paint street names on the curbs as students and children tend to steal the street name signs. They created a taxi stop in Govan Mbeki Drive, just outside the neighbourhood, and paved a path for children to ride their bicycles to school so they do not have to use the street. This initiative was in collaboration with the city council. One may interpret the fact that the taxi stop is located outside the neighbourhood in at least two different ways. It might have been placed there on a main route, for minimal disruption of the normal route the taxi driver serves. The drivers therefore need not pull into Oewersig to drop each and every domestic worker at their exact place of work. Alternatively, this might also have been a way to keep taxis and additional 'prying eyes' out of the neighbourhood. In this latter interpretation taxi drivers or occupants are seen as some sort of threat to security or for some even as a disruption to the exclusively white and upper-middleclass order of things.

A 'good neighbourhood'

Oewersig, meaning in sight of the river, is a leafy upper middleclass suburb. It is located close to arguably the most prestigious primary and high schools in the city and close to the North-West University. It is also a relatively secure neighbourhood. As explained, part of the reason for this relative safety is the street plan. By car the neighbourhood can only be accessed through two entry routes from the same main road that runs through a large part of the city. There is a pedestrian bridge over the Mooi River towards the high school's sports fields and residents have negotiated access to the adjacent university sports grounds through an access card-controlled gate. There they can enjoy walks in another picturesque setting. All of the residents interviewed mention reasons such as Oewersig being a 'sought after' neighbourhood, proximity to the aforementioned amenities and the closeness to nature as reasons for buying homes in the area and residing there for many years. They are seemingly in fact quite content. From the responses given by participants, home owners tend to buy property in this neighbourhood and stay for many years. One participant even noted that 'it was the best address in town' (Oewersig resident 3 Interview, 2018; Author's translation).

One might argue that upholding the status of the neighbourhood has been a major reason for the activities initiated in 2014. Having such a 'good' or 'nice' neighbourhood in post-apartheid South Africa is a notion in need of some deconstruction. But, this deconstruction I would suggest requires at least some caution, as we need to distinguish between legitimate concerns or at least reasonable concerns and less acceptable justifications for exclusionary practices. The inverse of good is of course bad and we need to ask ourselves what 'bad' means in this context and to these residents.

Firstly, South Africa is a capitalist country. Property rights, despite recent talks about expropriation without compensation, will remain enshrined in the South African Constitution for years to come. In keeping with the logic and purpose of this book, to not alienate elite groupings and think through the possibilities for potential alliances across diverse segments of society, this fact is taken as a given. Although it remained largely unsaid, at least explicitly, a concern with the value of their properties most likely influenced many of the activities of the residents' association, especially since the overwhelming majority of residents are home owners. Home ownership is a major investment. A bond is paid off over 20–30 years. It would make perfect sense for home owners to engage in activities they believe will make their homes more appealing to higher income buyers. Attaching labels such as safe and secure, well maintained and beautiful as well as proximity to schools and recreational facilities to the neighbourhood will surely help this cause. One participant reasoned along these lines by referencing property developers and other business people (Oewersig resident 4 Interview, 2018). She noted the following: 'What are the implications of crime for them?' "How do we keep the atmosphere of a middleclass neighbourhood where you can live with your children close to the university?"' (Author's translation).

Secondly, for some, the word 'bad' might imply dirty or filthy, unholy and inferior. The neighbourhood is next to the Mooi River. There are *braai* facilities next to the river along with many rubbish bins. Moreover, there were previously many invasive plant species that have overgrown the area. As mentioned, the association assures that the latter is regularly cleared. One participant however mentioned that 'people are pigs'. Outsiders who come into the area to relax next to the river, would dump their fast food containers on the grass instead of walking a few metres to the rubbish bin (Oewersig resident 5 Interview, 2018). Others would smoke marijuana next to the river. Regarding the lack of municipal upkeep of the area, including the cutting of grass by the river and adequate crime fighting this same participant stated that he refused to be pulled down (to the level of others) by politics. This indictment of the municipality may not be entirely unfair. However, if 'bad' is the level of others, there are questions to be asked about the aspersions cast upon these others. Having noted this comment, there does seem to be some internal tensions linked to this participant's position, which affirm the argument that people are complex and that individuals are never only one thing. He also noted profound sympathy for poorer residents of the Ikageng township, even stating that he is willing to pay a bit more and do some more work in Oewersig, such as serving on the residents' association committee, so that resources can be spent in Ikageng where he often did work. I was convinced that these were sincere and authentic comments. 'Those people have nothing' he noted (Author's translation) and other participants echoed this position (cf. Oewersig resident 1 Interview, 2018). We will revisit what at first glance appears to a contradiction in due course.

Another participant feared a type of 'creep of the unholy', if you like (Oewersig resident 4 Interview, 2018). There were things in town that bothered her, though not yet in the neighbourhood. Something had to be done to keep this 'consistent creep' at bay. In particular, the example of the Bult area was mentioned, where low density housing was gradually replaced with high density flat complexes. These new forms of housing are primarily aimed at students. Not only, is this area subjected to the more typical shenanigans of students, such as heavy drinking, noise and at times fighting, but there were no more resident home owners who took responsibility for the upkeep of their property. Moreover, flats are often rented to drug dealers. An informal discussion in 2012 with a senior policeman revealed that drug dealers tend to move on every three months to avoid capture. The words used by the aforementioned Oewersig resident were literally, 'we do not want to go the route of the Bult' and they want to maintain a neighbourhood where they 'can raise children' (Author's translation). For this participant upkeep seems to relate to dirt and the unsightly, while temporary residents are associated with the unholy. Both constitute the opposite of a 'good neighbourhood'. Compare the following quotation:

> There were questions asked about business people who want to start something in the neighbourhood. What are the implications of crime on them? We do not want to follow the same path as the Bult, where the people who concern themselves with the upkeep of the neighbourhood are worked out and now you have more temporary than permanent residents.
>
> (Author's translation)

It would also appear that a good neighbourhood is one with working infrastructure. Many residents complained about street lights that have either been switched off to save electricity or they are out of order (Oewersig resident 6 Interview, 2018; Oewersig resident 5 Interview, 2018; Oewersig resident 1 Interview, 2018). This has been repeatedly reported to the city council, but to no avail. Darkness, of course creates a potentially dangerous situation. It might provide an opportunity for criminals to evade detection, although this is possibly less likely with the plethora of cameras in the streets. The association is considering installing solar street lighting. Also batteries for each individual camera to provide power when the electricity is down. Another similar gripe concerns the matter of refuse removal. It has frequently happened that the Association had to take responsibility for removing refuse from the neighbourhood, because the city council was unable to do so. Having sat in CPF meetings, I know that a refuse removal truck that broke down had been out of service for long periods of time. More than one resident noted that they do not expect a lot from the municipality, but there are some basic functions, for which they pay every month that should be fulfilled by the municipality (Oewersig resident 3 Interview, 2018; Oewersig resident 6 Interview, 2018). Street lights and refuse removal are two such functions.

To summarise, it would appear that a good neighbourhood is safe and well kept. It has uninterrupted access to municipal services, such as street lights, refuse removal and the cutting of grass on public land. Its inverse, is what some appear to view as the potential for urban decay. Subsequent to the conclusion of interviews in Oewersig, municipal bylaws were drafted that allow for city improvement districts. One such district has already been declared in the area to the West of Oewersig and this CID might be viewed as a type of 'buffer' that might allay some of the aforementioned concerns. The CID is the topic of the next chapter. The remainder of this chapter brings the discussion thus far into conversation with the analytical framework of this book, in particular the very specific elements taken from Henri Lefebvre, namely notions of spaces of representation and representation of space and how the dialectic between the two is evidence of a fractured and potential opportunity laden hegemonic order as per the work of Laclau. The chapter will then conclude on the implications of Oewersig's security infrastructure for the rest of Potchefstroom and comparable settings.

Lefebvre's work on cities are meant to point out the 'contradictory mediation between social order and everyday life' (Kipfer, 2008:6). This can be compared to a more homogeneous industrial space. Heterogeneity necessitates constant articulation to clarify social reality (Prigge, 2008:50) or what Laclau would deem a sutured hegemonic order. But the totality according to Lefebvre is unpresentable (cited in Prigge, 2008:51). 'the thread has snapped between the real and the symbolic, between the existential experiences of everyday space and their representations in ideology, science, and culture' (Prigge, 2008:51). As such, we are left with the triadic dialectic explained in the introductory chapter to this book and outlined in *The production of space* (1991). This chapter is concerned with confrontations in everyday life as people 'seek to increase the(ir) respective room to manoeuvre and the(ir) articulation of interests by shifting the frontiers between dominant and dominated spaces' (Prigge, 2008:54). This dynamism links with the contestability of hegemony. What is at play here is the conflict between the (official) representations of space of the municipality and the unfolding post-apartheid social order and spaces of representation of the residents of Oewersig and the *de facto* spatial practices of the broader social order. The imaginary and physical reality of the laager are associated with spaces of representation. Such exclusionary spaces that serve the need of a minority are subject to the vacillating practices of semigration and recolonisation of public space described in Chapter 2. Residents in relatively powerful subject position are able to establish aesthetically pleasing to themselves and expanding defensible spaces, by successfully advocating for particular spaces of representation. Representations of space are *somewhat* more in tune with the symbolic and spatial practices or the 'real' for a broader segment of society.

Spaces of representation

Clearly, to talk about spaces of representation as per the work of Lefebvre the notion of place as opposed to merely space is integral. Some would like to see Oewersig become a gated community in the true sense of the word. This might not be that far-fetched as gated communities are increasing significantly on the edges of Potchefstroom. However, as it stands, Oewersig might be viewed as a *quasi-gated community*. Those who enter are virtually continually viewed on live camera, except for when they enter a house. What this quasi-gated community wants to keep in and out is not too different from real gated communities. Crime and security is a major concern and I would like to once again reiterate, a valid concern, in the context of South Africa. In addition, and again similar to gated communities, there are particular ways of living that residents want to preserve. These relate to cleanliness, tranquillity and an enjoyment of nature. Finally, this particular quasi-gated community is in one of the most sought-after parts of the city. Residents and in particular home owners want to preserve property values and a return on the major investment they made.

Participants in general agree that the cameras are working. The incident reports delivered by the PSC indicate that they (the PSC) are able to detect 'suspicious' behaviour and respond in a timely manner. However, most participants note that the cameras have given them peace of mind. It is not that they were previously overcome by fear. One participant noted in this regard that 'Die wat hier werk en woon voel veiliger' (Those who live and work here feel safe) and 'Die wat kom rondloop is meer versigtig' (Those who wander these streets are more cautious) (Oewersig resident 4 Interview, 2018). Others have noted that the initial shooting incident did not instil fear, but rather a determination to be proactive (Oewersig resident 2 Interview, 2018). There used to be a lot of drug dealing in the area. This has decreased significantly. In truth, it has likely been displaced onto other parts of the city. Some participants believe that new people move to the neighbourhood because of the security measures in place.

According to one participant, especially older people in the neighbourhood struggle with fear and as a member of the managing committee he is often contacted by these residents' every time something happens and it is reported on WhatsApp (Oewersig resident 3 Interview, 2018). For this reason, management developed the previously mentioned filtering protocol. Not all incidents are shared with the entire neighbourhood so that undue fear is not cultivated. One example of this is an instance of a man who was assaulted.

Although interviewees note that the cameras have been effective in reducing crime in Oewersig, there is still some crime. The most common forms of crime, are theft from people's yards or garages for example of lawnmowers or bicycles. Occasionally, thieves might enter the house. Hence, housebreaking remains a prevailing concern. Here the stories of apparently very 'professional' gangs operating in a neighbouring area causes much concern. In these instances, the dogs don't bark and there is a rumour that criminals use gas to neutralise the dogs.

These aspects constitute Oewersig as a space of representation for most residents. They are part of the economic elite. They might not all be part of the top 1%, but they all live upper middleclass lives. In order to ensure the many victories, they have achieved towards quasi-gated community status they were confronted by representations of space, often driven by a newer political elite in the form of the city council and its officials.

Representations of space: conflict and cooperation with the city council

The relationship between the residents of Oewersig and the city council has been rather chequered. Responses from participants range from 'there is no relationship with the municipality' (Oewersig resident 5 Interview, 2018) to more constructive and conciliatory attitudes (Oewersig resident 2 Interview, 2018). This is where the diversity within the neighbourhood and as such within a particular elite grouping becomes quite clear. As mentioned, some would like to completely close off the neighbourhood, while another participant noted that 'it's a pity that we need these cameras' (Oewersig resident 2 Interview, 2018; Author's translation). A small minority, especially a particular homeowner, was initially quite opposed to much of the project, citing concerns with privacy and freedom of movement. Consider the following quotation: 'One or two people had reservations about big brother watching me. This blew over very quickly' (Oewersig resident 1 Interview, 2018; Author's translation).

Also, it has been questioned whether the cameras work. Theft still takes place. The explanation is that this was in areas where there were 'blind spots'. And maybe the security company have missed one or two people entering the area. 'They are human, you look away and you might miss someone' (Oewersig resident 1 Interview, 2018). Indeed, I have been to that same control room and not all cameras are on screen at any given point in time. Furthermore, the space on the screen allotted to a particular signal from a specific camera is very small. Also questions can be raised as to the sustainability of the system. At the time writing the PSC charged the association a very low rate of only R2,300 per month. It is an open secret that monitoring cameras cost a lot. However, by installing these cameras in the streets across the entire neighbourhood, or close to it, residents become individual customers to the armed response services provided by the company. It is here where a loss turns into a seemingly significant profit, as the individual subscriptions far exceed the costs incurred by the PSC. However, the impoverished neighbouring township of Ikageng aside, will even lower middleclass suburbs in Potchefstroom be able to carry the additional levies per month for both the monitoring and armed response components? By my estimate the combined monthly rates should run up to R500–R550 per month per household. If this is not affordable, there may be implications for other parts of the city. The trend observed by a local PSC owner (cf. PSC owner interview, 2017), where that security improvements, simply shifts crime from one part of the city, seemingly the wealthier part, to other (less affluent) parts. This will likely continue. It is here where representations of space come in.

Continued segregation post-apartheid might not have been envisaged by many trying to work towards a new social compact from the early 1990s. Yet these unfortunate trends have been typical. Social justice and desegregation have been key objectives of many government officials and it is increasingly the case as city planners have designated part of town for low-cost housing projects, close to historically white and middleclass suburbs. Continued segregation potentially infringes upon the right to the city. This concept, introduced in Chapter 2 and conceived of by Lefebvre (1971), relates to equal opportunity for all to co-constitute the city and as Purcell (2013) argues, 'to live in it well'. Of course what exactly 'living well' means may be open to some interpretation. It is however adequately explained in a way that clearly brings it into stark contrast with the potentially oppressive social order. Lefebvre's position is based on an aversion to systems of thought that simplify urban realities (Lefebvre, 1971:63). These systems paint over the divisions and inequalities in a society by closing themselves off from the nuances of oppression in urban life. It appears that transformation of the urban life is meant to be achieved by pragmatic thought and action that identifies openings in everyday life for interventions, be they popular mobilisation or policy wise (Huchzermeyer, 2018). Consider the following quotation:

> In so far as there can be demonstration in such matters, we have demonstrated the non-closing of the circuit. There is no single, absolute chosen system but only sub-systems separated by cracks, gaps and lacunae; forms do not converge, they have no grip on the content and cannot reduce it permanently.
>
> (Lefebvre, 1971:63)

The right to the city can be a starting position or optic through which such openings are identified and acted upon. The objective of such intervention is not simply greater and broader access to public spaces, it also implies the ability of diverse members of society to shape public spaces, making them their own for a more broadly functional social order.

The Lefebvrian adherence to more open representations of space and a more progressive politics, is also reflected in a clear initial aversion to a gated community in Oewersig on the part of the city council. Moreover, such sentiments are partially found amongst the residents of Oewersig themselves. Consider the following statement: 'The PSC knows all the vehicles in the neighbourhood, but do you want to be known like that?' (Oewersig resident 2 Interview, 2018; Author's translation). Although more conciliatory dispositions might be in the minority, they do exist and importantly, they do sometimes correlate with questions raised by municipal officials. Yet this is the same city council (be it not a homogenous entity) with which a taxi stop outside of the neighbourhood was negotiated in exchange for the residents' association paying for a fence elsewhere in town.

It seems clear that in the dialectic between spaces of representation and representations of space the former and therefore the residents of Oewersig have by and large won. They gained far more than they have lost from their interaction with the city council. Yet, one has to have some sympathy with Oewersig residents. They have the right to take responsibility for the environment they would like to live in and to protect the investment they made in a rather expensive neighbourhood. Moreover, it is not just the middleclass that is dissatisfied with municipal service delivery. I have numerous found complaints about similar matters throughout Potchefstroom and Ikageng.

Critical reflection

The gap between state and society, mentioned throughout this book, has facilitated some logical leaps and persecutory perceptions. This attack is not only from fellow citizens, but also from a city council often voted in by non-income tax payers and – it is supposed – the same cohort from which criminals are drawn. The SAPS are at times viewed as serving others, often criminal interests and being extremely corrupt. While corruption in the SAPS throughout South Africa has been repeatedly reported and even acknowledged by the service itself (cf. Newham and Rappert, 2018:8; Samara, 2011:33; Altbeker, 2005:180, 187), some of the conclusions drawn by certain participants are questionable. Consider the following: 'The municipality does is not interested in white neighbourhoods' (Oewersig resident 5 Interview, 2018).

Comments such as these are troubling. They help to constitute the laager as a central piece of security infrastructure. Similarly, social media and interviewees from Oewersig report how the reaction by the PSC is extremely effective. They would view something or someone suspicious and immediately send out an armed response vehicle as a proactive measure. The presence of such a major piece of infrastructure also enables comments such as the following. 'The people who just walk around the neighbourhood looking for trouble are far fewer now' (Oewersig resident 3 Interview, 2018; Author's translation). The troubling part is not the person walking in the street. The troubling part is the lengths to which the response to criminal activity by a small minority has gone and how security infrastructures are in a symbiotic relationship with toxic dispositions vis-à-vis fellow citizens. Dismissing fears of crime would be irresponsible and foolish. However, the case of Oewersig demonstrates the *de facto* effect of expanding the physical laager, even though this might not have been the intention. As one participant duly noted: 'This model is possibly the friendliest. We want to be an open neighbourhood' (Oewersig resident 4 Interview, 2018; Author's translation).

There is merit to this characterisation as management and the PSC take much care to treat data ethically and to ensure that it is only used for its intended purpose. A member of management for example shared a video of a girl in school walking past a car and exchanging something with the driver. So too were stillshots of the occurrence. I was requested to not publish any of this material. This was most likely a drug deal and there have been others like it. The association shared the incident with the nearby school to make them aware of possible drug use amongst their students, but neither the residents' association nor the PSC followed up on this and similar events themselves. They did share the (presumable drug dealer's) vehicle description and registration number with the SAPS.

The point of the cameras is to allay fear of the most immediately severe losses of property and threats to human life and limb. That is the basis of the agreement with the PSC. Some misdeeds are intentionally treated confidentially. This might be smart, as a more intrusive approach could lead to less buy-in from the public in Oewersig and other parts of the city. Be that as it may, it limits the feeling that everyone is constantly policed even though they are, in Oewersig, constantly under surveillance. And this is of course a surveillance as we have discussed previously that is unequally applied to different segments of society.

In this regard one participant cited earlier noted that he would have wanted a different type of social order where 'he fits into the public'. This participant, who is an elder in the church, noted that even when he goes to visit some residents on church business, that it is difficult to access their homes, as they don't even have doorbells anymore. They only have high walls around their houses. He notes that: 'There is something healthy about it when you are able to leave your front gate and you are in public and that you are open to people from outside' (Oewersig resident 2 Interview, 2018, Author's translation).

So how do we attempt to deal with this complex situation. The concluding section draws on the work of Laclau to briefly highlight some potential constructive points of interference. A key aspect in this regard is the fact that no category of person, including residents of Oewersig are homogenous in every single way, including what and how they think.

Conclusion

There are particular dangers in the fact that bylaws are being finalised and that in theory any neighbourhood could soon be able to declare a non-profit company, a special rates area and a CID. These are the dangers of insularity in a social order that has struggled with social cohesion for its entire history. Infrastructures that facilitate insularity include the laager and constitutive elements such as fibre-based internet, cameras, PSCs and the horse shoe-like street plan of this particular neighbourhood. All of these tools work together to produce a particular politics of crime prevention whereby signifiers such as 'victimhood' of the neighbourhood (a form of securitisation), 'the good neighbourhood' and the 'creep of social decay' and resultant actions circulate. Even if this is not always intended (which I am convinced) the combination of an infrastructural substrate, divisive signifiers and security practices do serve particular hegemonic interests. The conclusion of this chapter is not, and cannot be, that Oewersig immediately reverse a set of steps that have clearly given residents significant peace of mind or ontological security. We should, however, ask a few pertinent questions. Firstly, are there some perceptions amongst participants that are potentially problematic and in need of constructive engagement? The answer is 'yes'. In the bigger scheme of South Africa's political economy and physical (in)security, the residents of Oewersig are not the primary victims and comments to that effect reveal an alarming perspective, which only drives further division.

Secondly, and arguably more importantly, is the precedent set by the case of Oewersig the way we want the entire city and other parts of South Africa to go? The answer is surely 'no'. This is not constructive in dealing with the perennial reiteration, as opposed to reproduction, of the country's divisive past. However, for meaningful intervention in this matter the government has to empower the state to be better prepared to fight crime, to stimulate greater job creation and at the level of the city council, to organise more constructive interactions between citizens across all lines of division. These might be more informal events, not characterised by accusations (which might have merit), but where there is opportunity for greater sympathetic discussions between diverse South Africans. Finally, my aforementioned choice of 'reiteration' above 'reproduction' points to the very nature of hegemony. It is fragile and subject to consistent change. Therefore, the lack of coherence in elite and other groups can be used as means to establish conversations between 'the right' people from diverse groupings when the aforementioned opportunities for constructive engagement are planned. Change may at times be subtle, but it is always imminent. It should however be steered, consistently, in the right general direction.

Notes

1   *Oewersig* may be directly translated as within sight of the river bed. More loosely translated we might take the name to mean, the neighbourhood located on or next to the river. The river in question is the *Mooi* (meaning beautiful) River. This is an upmarket neighbourhood. Many professionals, local business owners and professors own homes in this neighbourhood. Most of the academics who bought houses in the neighbourhood are older and bought their homes quite some time ago in the first five to seven years of the current millennium. It is unlikely a professor, even a full professor, in a single-income household would be able to comfortably afford a house in this neighbourhood today. Oewersig is a sought-after address in the city.

2   On 6 January 2020 the exchange rates were as follows. ZAR 12,000 = USD 893 and ZAR 2,500 = USD 175 or EU 751 and EU 156.

3   The Cachet Park CID, which is the topic of the following chapter, was subsequently formed. At the time of writing bylaws, which govern such an arrangement had been drafted, but were not yet approved by the City Council.

References

Altbeker, A. 2005. A funny thing happened on the way to an integrated justice system. *South African Review of Sociology*, 36(2), pp. 178–190.

Clarno, A. 2013. Rescaling white space in post-apartheid Johannesburg. *Antipode: A Radical Journal of Geography*, 45(5), pp. 1190–1212.

Huchzermeyer, M. 2018. The legal meaning of Lefebvre's right to the city: Addressing the gap between global campaign and scholarly debate. *GeoJournal*, 83(3), pp. 631–644.

Kipfer, 2008. How Lefebvre urbanized Gramsci: Hegemony, everyday life, and difference, in K. Goonewardena, S. Kipfer, R. Milgrom and C. Schmid (eds.) *Space, difference and everyday life: Reading Henri Lefebvre*. London: Routledge.

Lefebvre, H.1991. *The production of space.* Translated by Donald Nicholson-Smith. Oxford: Blackwell.

Lefebvre, H. 1971. *Writings on cities.* Translated by Eleonore Kofman and Elizabeth Lebas. Malden: Blackwell.

McCain, N. 2018. Special rating area 'the way to go' in Peoples' Post. Online: www.news24.com/SouthAfrica/Local/Peoples-Post/special-rating-area-the-way-to-go-20180611. Date of access: 8 January 2020.

Newham, G and Rappert, B. 2018. Policing for impact: Is South Africa ready for evidence based policing. *South African Crime Quarterly*, 64, pp. 7–16.

Prigge, W. 2008. Reading the urban revolution: Space and representation, in K. Goonewardena, S. Kipfer, R. Milgrom and C. Schmid (eds.) *Space, difference and everyday life: Reading Henri Lefebvre.* London: Routledge.

Purcell, M. 2013. To inhabit well: Counterhegemonic movements and the right to the city. *Urban Geography*, 34(4), pp. 560–574.

Samara, T.R. 2011. *Cape Town after apartheid: Crime and governance in the divided city.* London: University of Minnesota Press.

Interviews

Oewersig resident 1.2018. Interviewed by Gideon van Riet. 8 November. Potchefstroom.

Oewersig resident 2.2018. Interviewed by Gideon van Riet. 7 November. Potchefstroom.

Oewersig resident 3.2018. Interviewed by Gideon van Riet. 13 November. Potchefstroom.

Oewersig resident 4.2018. Interviewed by Gideon van Riet. 15 November. Potchefstroom.

Oewersig resident 5.2018. Interviewed by Gideon van Riet. 19 November. Potchefstroom.

Oewersig resident 6.2018. Interviewed by Gideon van Riet. 19 November. Potchefstroom.

Private security company owner. 2017. Interviewed by Gideon van Riet. August. Klerksdorp.

9   The Cachet Park City Improvement District

Introduction

Various aspects to be dealt with in this chapter have been introduced briefly in previous chapters. Chapter 3 dealt with the contradictions of frontier governance in the greater JB Marks municipality, in light of significant historical entanglements. This chapter deals with the Cachet Park CID, in historical perspective. There were previously similar though far less comprehensive interventions in that part of the city. The analysis will start with those and concerns over crime in the Bult area in recent years. Chapter 8 dealt with the issue of voluntary surveillance bought from a PSC by a residents' association. The CID has benefitted from the same technological advances to set up a system of surveillance to which the neighbourhood of Oewersig pales in comparison. Surveillance on this scale is a cost-effective approach to policing a significant part of the city.

Chapter 1, in addition to explaining the logics of closure through abandonment and frontier governance, offered a particular interpretation of crime in Potchefstroom. Accordingly, crime is linked to extreme inequality in a context of equally extreme unemployment, multidimensional poverty, and senses of injustice after democratisation. This is not a comprehensive account of crime. There can likely be no such thing. But, for reasons explained in that chapter, crime cannot be understood outside of the dynamics mentioned earlier. Such a reading of crime has served as a benchmark through which the response to crime in Potchefstroom, mostly outside of the state, has been critiqued. The CID too makes use of a suite of security infrastructures and associated practices of target hardening. In addition, there is a major risk, based on a review of CIDs as a phenomenon and evidence from Potchefstroom, that this set of infrastructures could reiterate post-apartheid laager conceptualised in Chapter 2 and effectively police perceived difference. As such, similar to the case of Oewersig discussed in Chapter 8, there is potential confluence of particular paces of representation, scalar fixes, semigration and the recolonising of public land to which the state and political elites offer little resistance. The policing of difference does not only go against the principles of a constitutional democracy, but it appears to have major implications for some of the basic livelihood strategies of the urban poor, that have been dependent on this area. As a counter to these strategies the chapter refers back to the perspective of Chapter 1 on crime in South Africa. The chapter makes an argument for a return

DOI: 10.4324/9781003028185-13

to a more holistic approach to crime and as such additional funding to social programmes. This strand of thought will be developed further in the concluding chapter. For now, it can be stated that a holistic approach to crime, requires not only
funding to diverse departments, but also an end or at least a major arrest of the
toxicity of local level politicisation of the bureaucracy. At the national level, political will for macroeconomic policy execution that intervenes upon the problem of
unemployment is similarly indispensable. None of these suggests that crime and
fear of crime are not valid concerns. The argument is simply that the response to
crime documented in this chapter and the book in general will likely fuel division
and in the end not sufficiently deal with the problem of crime.

Methodologically this chapter draws on key informant interviews, attendance
of a Sector Four CPF meeting, the public Facebook page of the CID, field notes
from direct observation in the Bult area before the formation of the CID and newspaper articles. These sources are brought into conversation with the theoretical
literature on CIDs, and Laclauian interpretations of hegemony. The latter implies
that key livelihood options of the urban poor are threatened and that crime is
potentially displaced onto other communities.

The remainder of this chapter is structured as follows. The following section
provides an overview of the area in question, its recent problems with crime and
the recent history of interventions. The section will explain what has led to the
formation of the CID. Thereafter the chapter offers a brief literature review on
CIDs. These infrastructures have a chequered history. They often entail the harassment and policing of difference and the displacement, instead of the reduction
of crime. That section will also give a general overview of the Cachet Park CID
and its main functions. Thereafter the chapter deals with the substantive practices
of the CID and the implications of these activities. Some of this discussion will
be grounded in existing evidence and some will be more conjectural, though well
motivated given the literature review on CIDs and evidence documented in this
and previous chapters. Again, it must be emphasised that this chapter and this
book in general is not meant to vilify specific actors. Although I often disagree
with many actions emanating from security infrastructures, the analysis should
be ethical. This implies a level of understanding and respect for those who are
researched, which, in turn, requires a good faith attempt to appreciate the reasons and thought processes that inform security practices. There is much need to
reduce fear, which will necessarily require a reduction in crime. I would like for
my analysis to be understood as a critique at the level of practice; in other words,
the means by which security actors seek to intervene upon the problematic of
crime. In addition, the purpose of this book is to also move beyond critique, or at
least polemic, by making some tentative suggestions for practice. The chapter will
conclude by considering once more through a Laclauian lens, the prospects for
democratising the politics around the CID in a way that does not decrease safety
throughout the city and does not lose the buy-in of key stakeholders.

The central argument of this chapter is that CIDs have generally been problematic
in the past. The Cachet Park CID is still relatively young and can learn from the
adverse consequences of other CIDs. The main way to do this is to explore modes of
thought and resulting practices, where the security of some is not viewed as dependent

on the exclusion and insecurity of some of the most vulnerable and impoverished in society. Moving towards a radical and plural democratic social order, including productive logics of construction, will require a new politics of security in the face of crime, where the aforementioned values are upheld. In other words, the hegemonic order may be rearticulated in two ways. The first is towards exclusionary and conservative ends. The second is towards progressive and inclusive ends. The latter, especially in the long run is better for democracy and crime reduction. In light of a government that is increasingly ineffective, yet unlikely to be voted out of power, it is imperative that the people and private and semi-private institutions articulate solidarity across reiterated fault lines is into the dynamic status quo.

The area in question

The CID covers a limited number of street blocks, roughly between Thabo Mbeki Drive to the East and about three street blocks West of the railway line between Cape Town and Johannesburg to the West. At its heart is the Bult area. This place name is a misnomer. Directly translated Bult means hill. This part of the city, much like the rest of Potchefstroom has very little topographic features. The Bult,

Figure 9.1 Photo of Cachet Park: taken from the South-West

centres around Cachet Park, which is both the name of a park and a small shopping centre immediately to the North of the park, which is owned by the university. The park covers an entire street block. The shopping centre to its North includes a supermarket, a pharmacy, a print shop, a laundry math, a book store, an estate agency, some restaurants and a gymnasium.

To the South, the park is flanked by a small retail block, a motel and a block of flats. To the West, along Hoffman Street, the park is flanked by university hostels and the entrance to the administrative head office that oversees all three NWU campuses. The university is shielded from the rest of the city through a palisade fence. A total of six gates give pedestrian, motor vehicle and bicycle access to the campus. To the East, there is a parking lot that caters to the various retail outlets, restaurants, bars and nightclubs across Steve Biko Street. These establishments in turn cater mostly to students and include another book store, coffee shops, convenience stores (called cafés in the South African context), fast food vendors, a bank, liquor stores, a tattoo parlour, venues that host live music, nightclubs and other 'watering holes'. From this street there is an alleyway that leads to some of the aforementioned businesses.

Previously there were informal initiatives to reduce crime on the Bult instituted by the ward councillor for this area. Much like the initiatives under the CID, discussed in the following, this initiative took car guards as a major source of crime in the area. They were reportedly involved in drug dealing and stolen goods sales (Interview, car guard, May 2018; Interview, shop owner on the Bult, June 2018). Other crimes in this part of the city included rape and theft of motor vehicles and the aforementioned attacks on students (Saayman, 2019:3; Saayman, 2018:2; Wetdewich, 2018:5; Potchefstroom CPF Facebook group, 10 September, 2017). 'Project Clean-up' involved a set of uniformed 'trusted' car guards. The public were encouraged to only support these individuals. In addition, these car guards were tasked with keeping the area around the park clean. Later developments in crime reduction include both the CID and prior to this the subdivision of CPF Sector One.

Sector Four came about following discussions between the head of Campus Protection Services (CPS) and the Potchefstroom police station commander. These men used to be colleagues within the SAPS. They were concerned about the manifold crime problems they faced and how to deal with those 'structurally'. They called a general meeting with key stakeholders, such as the councillor for this part of the city, the university and home and business owners. In this meeting a sector profile was developed. Subsequently positions were advertised and office bearers were nominated and elected. This structure, like any other CPF sector has monthly meetings where crime related issues are discussed. In this meeting the CID, subsequent to its formation, is also represented. Somewhat different to other sectors, this meeting draws on two sources of crime statistics. The SAPS provide crime statistics, while CPS provide statistics on crimes committed on campus.

Every Monday CPS have an 'extended crime combatting forum'. Here they evaluate the effectiveness of risk control measures. In that meeting the sector

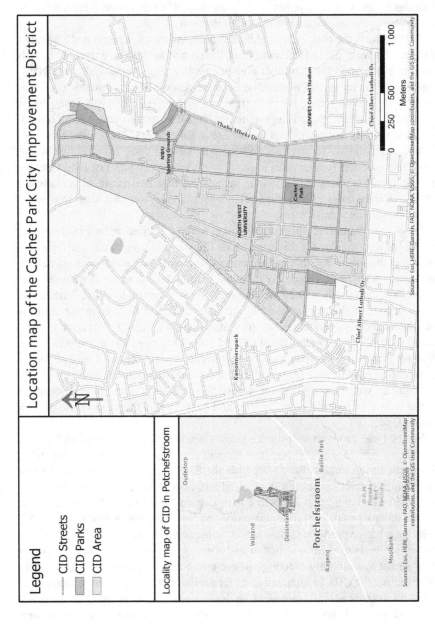

Map 9.1 Location map of the Cachet Park City Improvement District

commander is also presented, as well as the CID and a PSC that provides 'static deployments' on the campus. As explained to me by the head of CPS, this second PSC looks after 'referent objects of security' in the Cachet Park shopping centre (Interview, CPS Head, February 2020). They also serve as 'force multipliers' at the access control points to the university campus. In addition, there is a parade every day at 05:30 and 17:30 where CPS officers share information about suspicious motor vehicles, persons and crime trends and responses. Hence deployments by CPS are not haphazard. They are based on evidence.

There are clearly overlapping partnerships for crime prevention in this part of the city. The CID is part of the somewhat larger Sector Four. There is cooperation or at least an understanding between the SAPS and the university through this particular CPF. There is also overlap in terms of the stakeholders involved in both structures. It is not just the one PSC that maintains most of the CCTV cameras in the city, that is involved in this part of the city. There is also a second PSC, which is used to augment the work of CPS.

City improvement districts as a phenomenon

City improvement districts in South Africa started in Rosebank in Johannesburg in the 1980s. In the literature, CIDs are treated as part of enclavism (cf. Wissink, 2013). As such, they are not so much a new phenomenon, as they are a reiteration of existing logics. Consequently, Hannigan (1995) questions to what extent enclaves such as CIDs are really leading to a new urbanism. Authors such as Wissink (2013:2) warns against the presumptive 'narrative of loss' and 'alarmist studies of fragmented cities in the post-industrial economy' in existing literature on CIDs. These accounts, apparently, do not recognise the benefit of some forms of enclavism for disadvantaged communities, for example in freeing up state capacity for delivering public goods (also compare Schuermans, 2016:183). McKenzie (2011) argues that it is possible to limit the detrimental effects of enclavism. In the shorter term this is most likely the logical objective in South Africa. Others such as Schuermans (2016:184) agree that there is a narrow focus on social justice, crime displacement and limited social solidarity through a focus on the residential site of encounter only. Stated differently, interaction between different groups also occurs outside of the CID/enclave. This is unavoidable. However, critiques such as that of Wissink, I would argue, are less robust in South Africa's context of historical segregation and incessant border-making. Also, what I am referring to is the continued practice of various degrees of informal border-making often based on a continued logic of apartheid, be it race, class or otherwise and more complexly based. Unlike authors rightly critiqued by Wissink, I am not invoking a notion of a 'previously integrated city', nor am I arguing that enclavism has ever or does today mean complete separation. Also these studies, cited by Wissink, assume a decline in social cohesion. My argument, in the context of Potchefstroom, is that the starting point is a relative lack of cohesion and that enclavism does not help.

This book, as an analysis of a smaller city and not focussed on gated communities seeks to contribute to an understanding of the diversity between enclaves. Some link the growth of these districts to neoliberalism and its property ownership regimes (Didier et al., 2012:915). These authors hold the narrative of a North-American neoliberal model exported and accepted by those with influence in South Africa. I am, however, not sure about the completeness of that narrative. There is something about the South African context and certain local imaginaries that made some yearn for such a form of organisation. City improvement districts in South Africa, though related to capitalist agendas, is also about fear, not just of crime, but also fear of change. There are often racist, in addition to classist overtones to these arrangements. Moreover, we should be careful not to perpetuate the notion of a supposedly coherent neoliberalism that expands across the world and conquers all. Instead, analysts should offer clarity on which elements of this broad rhizome is at play and why this is problematic. That being said, property ownership regimes give a type of moral authority to PSCs (Samara, 2010:647). Even owners are policed through strict membership contribution requirements, where the correct municipal by-laws are in place.

Others do argue the modes of governing employed through enclaves are not entirely reducible to neoliberalism (Lippert, 2014). Services provided even extend as far as employing its own social workers and close ties with local shelters (Berg, 2013). However, these additional social responsibility services might be linked to the maintenance of property regimes and exclusionary practices in general. Social services, in this sense, are offered elsewhere and thus help maintain the enclave. Rink and Gamedze (2016) usefully augment the neoliberal interpretation of CIDs by focussing on mobilities. They argue that CIDs enable the mobility of some and apply friction to the mobility of others, typically those deemed superfluous to the needs of the economy, as defined by those in positions of relative power. The very notion of 'improvement' according to these authors is premised upon the latter (Rink and Gamedze, 2016:654). The work of Rink and Gamedza is quite in tune with approach taken throughout this book. Paasche et al. (2014:1560) neatly sum up this link when they state that 'work of previous researchers often focusses on "preventing the worse" through the creation of literal and metaphorical barriers to keep out ideas, things and people that are seen to threaten the security of these spaces'. Such understandings are clearly easily reconcilable with the physical and ontological structures which comprise the laager defined in Chapter 2. The 'security of these spaces', clearly also relates to the ontological security of privilege, described in Chapter 6.

South Africa over the years adopted the North-American model of business improvement districts and expanded its usage, under the label CID, in diverse settings, including suburbs (cf. Samara, 2010:643). Nearly all of this research is focussed on the large metropoles, especially Cape Town, and some secondary cities. I am not aware of any studies in cities with populations below 500,000. South African CIDs are typically based on the logic of public–private partnership, where the argument is held that services within cities have declined and need the private sector's involvement (Didier, 2012:921). With declines in property values in city centres, tax income also declined. As self-funding units, CIDs therefore provided a welcome injection of revenue to cities (Didier et al., 2012:924). Here the politicisation of the bureaucracy is often contrasted with the 'pure' intensions of the private sector (Didier et al., 2012:929). Some have argued that CIDs have become tools for reiterated racial governance (Samara, 2010:640) often through the substitution of notions of threat to property for a threat to the apartheid racial order (Kempa and Singh, 2008). Here we find 'a politics of belonging organized around consumption, leisure and recreation' (Samara, 2010:644). City improvement districts in South Africa have often taken a zero-tolerance approach to crime, borne out amongst other ways in how car guards and homeless people have been forced out of these areas (Didier et al., 2012:929).

In one of these studies Schuermans (2016:183) makes the insightful observation that the objectives, as deduced through the actions of CIDs, is often not merely crime prevention, but a type of escapism from observing the impoverished black other. In this regard Samara (2010:640) talks about the rise of the panopticon city and the rise of the fantasy city. The fantasy city is one of selective consumption of Europeanised architecture and consumer culture, very similar to semigration. In the context of this chapter semigration has less to do with architecture than. As we will see, it has with particular cultural practices. The panopticon city is one of surveillance, policing and one might add, logically, frontier governance, and friction applied to the mobility of some and not others. (Clarno, 2013:1192). There are, however, limits to such escapism (Schuermans, 2016:190).

Lemanski (2006) makes a similar argument that high levels of fear amongst white South Africans cannot be reduced to fear of crime only. Instead, she coined the term 'fear of crime plus'. This observation is consistent with my arguments around reiterations of border-making through frontier governance. It is mostly the economic elite that drives these processes, although with limited resistance by the political elite. Some have argued that more interaction between diverse groups do little to reduce deeply rooted racist or prejudicial beliefs (Lawson and Elwood, 2014; Valentine, 2008; Wilson, 2011). Such an observation is problematic given the dynamics of party politics in South Africa, with which I will engage more deeply in the final analysis in the concluding chapter. It seems unlikely that a better government will be voted into power or that major reforms within government will be possible without pressure through strategic alliances across the reiterated fault lines of the social order. Moreover, if we accept this conclusion then it leaves very little by way of options for nation-building. Other scholars have, however, argued that spaces of encounter are potentially productive in challenging continued racist and classist beliefs (Besteman, 2008; Broadbridge, 2001).

Some observations within the literature on CIDs speak to the search for hope in light of the lack of homogeneity within groups this book is concerned with.

Significant to the Cachet Park CID and Potchefstroom in general is a frequent frustration observed and acted upon by some CIDs with scattered litter in the mornings after refuse was left outside over night for collection (Schuermans, 2016:188). The 'sakskeerder' (bag shredder) discourse has been prevalent in Potchefstroom and likely informs some recycling initiatives. Yet, some residents of CIDs would be concerned by the poverty they observe. Some would pack recyclable materials and leftover food in separate bags. Others would give money to passers-by against the wishes of the CIDs governing structures (Schuermans, 2016:188). When viewed through a Laclauian lens of an unstable sutured social order the contradictory observations mentioned in the previous paragraph become intelligible as normal manifestations of social orders and subjectivities that are complex as opposed to neatly bounded. These tensions should be viewed as potentially productive. Admittedly, the evidence, based on various case studies, is clearly that these are often largely exclusive spaces from which unwanted people are often removed and where homeless people are harassed.

The Cachet Park CID

The Cachet Park CID became operational in January 2019. The idea came about after a visit by university staff to the University of Pretoria (UP). At UP, a CID was considered helpful as accommodation on campus is limited and students live on the periphery of the campus. Students have to work in computer labs at night and the equipment they use, cell phones and laptops are targeted by criminals. Establishing a structure that directs additional security measures around campus and student accommodation would hopefully reduce robberies. As The NWU's CPS only has a mandate within campus, this institution had insufficient reach over areas where crimes against students often take place. In addition to the UP, the University of Cape Town (UCT) is also part of a CID. The NWU decided to copy this model and engage stakeholders around campus to this end.[1]

The CID is a not-for-profit company. The university provided start-up capital for the project. The CID has eight directors. Three of these are from the NWU. The remainder of the directors are other stakeholders, such as owners of buildings and businesses off-campus. Directors meet quarterly. As for daily operations, the following full-time staff are employed directly by the CID. The chief executive officer (CEO) is a retired NWU professor. The chief operations officer (COO) is a retired military brigadier. Operations include security, landscaping and cleaning. All these services, including a large-scale recycling initiative, have been outsourced. A service provider collects recyclable materials every week from designated drop-off points. Security, is dealt with through a PSC and CPS, which as mentioned only operates on campus, as per its mandate. Security interventions introduced by the contracted PSC include foot patrollers, an auxiliary incident reporting centre on the edge of the park, staffed by the PSC, a dedicated response vehicle for the CID and significantly well over a hundred CCTV cameras monitored remotely 24 hours per day.

Figure 9.2 Photo of the auxiliary incident reporting centre: taken from the West of Cachet Park

These cameras are also connected to facial recognition and number plate recognition software, but according to interview data not to national databases. Therefore, such recognition is only in relation to previous observations in the area. Another service provider does landscaping in the park, 'so that students can go sit there under the trees in a safe and secure environment' (Interview, CPS Head, 2020). The CID through, service providers, manages above ground services, such as cutting trees in the area. The municipality provide fundamental infrastructure, such as water, sanitation and electricity.

Other staff directly appointed by the CID includes a finance officer, a media officer and a recruiter. The latter person recruits new members who pay a monthly fee based on the value of their property. Individual contributions are a proportion of the value of the property compared to the entire area of the CID, which is valued at around ZAR 4.5 billion. As such, an individual contribution is ZAR R 225 per month is expected for every ZAR 1 million in property value (Cachet

Park Facebook page, 21 June 2019). I have seen the draft bylaws, which were kindly shared with me by the municipality. They struck me as extremely vague and nowhere near as strict and enforceable as reports on similar arrangements in larger cities have suggested. Much like the Oewersig Residents' Association discussed in Chapter 8, the Cachet Park CID, in light of incomplete bylaws, uses a recruiter and frequent social media posts demonstrating success and membership levels, to convince others to contribute.

Unfolding substantive practices: Many, many more cameras in the street

The CID boasts far over a hundred CCTV cameras. For the most part these are installed on pylons erected on street corners. Four cameras are attached to each pylon, each camera facing into a different street. These cameras are monitored by the PSC 24 hours per day.

Figure 9.3 Photo of CCTV cameras in the Cachet Park CID

Based on previous research on cameras in public spaces Sandhu and Haggerty (2016:101) criticise what they call the common-sense assumption that cameras aimed at public spaces would reduce crime. Such interventions, according to these authors, are not evidence based. Black men draw more attention (Sandhu and Haggerty, 2016:102). Based on my research in Potchefstroom and the literature reviewed in this book, such a racialised gaze seems to be a typical consequence of private policing. It may be, because some within positions of power view race, qua race, as suspicious or it may be because of the significant overlap between race and class. As I have mentioned previously (Van Riet, 2020), the racial gaze is not always the fault of the PSC. They are beholden onto clients in an extremely saturated South African market and individual clients have often required such racialised policing. This problematic offers an opportunity to the CID to distinguish itself from a long history of racial and classist policing in CIDs, by openly framing itself differently. It is too soon to tell exactly the direction the CID is going. Breaking with such forms of policing will possibly place an inordinate training burden on the PSC in question, in a time when these types of initiatives are taking off at a rate of knots. Furthermore, as Sandhu and Haggerty (2016:102) note, manning the screens in control rooms to which camera feeds are streamed, is tedious work compounded by the fact that screens typically display too much information. Whether imparting the skills and up to date intelligence upon these operators is cost effective is an open question. We may expect of the CID as a major client to demand a type of policing more in line with the democratic principles described in Chapter 7. The CID can either draw on examples from elsewhere and rearticulate local politics in divisive ways or it can openly break with those examples by outright taking a position contra the observation cited earlier, whereby 'improvement' is viewed as such because of exclusion. Abandoning and banishing a vast and increasing segment of society cannot be viewed as improvement. A key question to ask would therefore be, how do we reduce crime and fear, without inflicting violence (broadly defined) upon others and without articulating additional fractures into the South African social order? Given the manifold crises plaguing South Africa and the politics between its citizens (Van Riet, 2020; Hart, 2013), this is a challenge that must be accepted. Unfortunately, there is evidence of policing and other initiatives in the CID taking on familiar and unfortunate forms, with a local inflection. These are discussed in the following sections. Immediately in the following, the discussion reports on the perceived success of this technology from within the CID board.

Those who are involved in the CID believe that it has made a major difference in terms of security and cleanliness. Statistics indicate a reduction in crime within the CID (PSC manager e-mail, 2021; CPS Head, 2020). The explanation given is that the opportunities for crime (precipitating factors) within the CID have been reduced as technology (i.e. cameras) have made potential perpetrators aware that they are being observed. It is however difficult to comment on predisposing factors. Because predisposing factors are so difficult to control, much of the previous literature on CIDs argues that CIDs merely displace crime onto other communities. This prospect is more significant since vast parts of Sector

Four and surrounding areas are taking on the model of Oewersig and the CID, that of live monitored CCTV cameras in public spaces. If one considers viable opportunities for crime, then there will likely be a displacement of some forms of crime towards the South of Potchefstroom and possibly into the Ikageng precinct. The large organised house robbery gangs, briefly mentioned in the previous chapter might start to avoid Potchefstroom altogether. These criminal activities are, however, not the most common forms of crime in the city. Everyday petty theft, housebreaking, some house robberies and stolen goods sales will likely move elsewhere within the city. Drug dealing, given the market in a city which has various higher education institutions, will likely also move to other parts of the city. There have over the years been a consistent decline in police officers per capita in South Africa (Gerber, 2018). Consequently, increased arrests and convictions that would imply crime eradication as opposed to displacement seem quite unlikely. Based on the discussions on the SAPS' capacity in previous chapters, it would appear that the potentially positive consequence of the CID freeing up capacity for the rest of the precinct will be limited at best. One alternative to displacement could be more brazen, violent and armed forms of crime in the same areas. If we take this view, then 'target hardening' should be understood in relative terms. The security of individual business or homes may be viewed relative to other suburbs, the city centre, the Ikageng precinct or even relative to a neighbour. Displacement can therefore be considered at different scales.

Crimes associated with significant levels of desperation are likely to continue. The most pertinent question appears to be where these crimes will take place. Previous studies on CIDs have also warned against the displacement of public policing to a situation where private security, previously only focussed on petty offences and by-law infringement, becomes a first port of call (cf. Paasche et al., 2014). As previous discussions in this book have demonstrated, this has been the reality in Potchefstroom long before the CID and residents' associations came into being. Forms of displacement in Potchefstroom have been taking place for quite some time. Residents' associations, CIDs and associated by-laws potentially offer more quasi-legal means by which actors may facilitate further displacement. There can be little doubt that security in the Bult area had become a major concern in recent years. As such, the local businesses, drawing on students, the revenue to the city council from these establishments and of course the security of students and other workers, owners and residents in the area were increasingly threatened. Increased security measures are therefore not unreasonable. Given the reasons provided here, they seem quite rational. What we may question and seek to engage with constructively, is how to weigh up of the interests of different stakeholders, including homeless people, car guards and other groups already mentioned in this chapter. This is not easy and critiques of this analysis should be welcomed for the sake of constructive intervention. The critique that will follow is pragmatic, though informed by the overarching argument that greater and meaningful interaction between conflicting groups seems increasingly to be the most likely source of progressive change in terms of effective crime reduction and security, amongst other concerns, within the South African social order. If a more holistic approach to crime is needed, then more effective governance of all the contributing dimensions of crime reduction is needed. Regardless of who the government is, pressure from broader and less immediately self-interested alliances will be required. This in turn will require more inclusive and collaborative politics from those in positions of power directing security flows through infrastructures such as CIDs and residents' associations in ways that are neither domineering nor paternalistic. It should be clear by now that this is not a call for 'politicisation'. What I am proposing is the reshaping of politics, an omnipresent activity, towards more productive ends.

Precarious livelihoods

City improvement districts as an institutional form has historically had adverse consequences. There is some evidence that the Cachet Park CID might be exhibiting some of those adverse consequences. However, some of the individuals involved in developing this CID have had foresight and have done admirable work in mitigating some of the potentially adverse effects of the CID. The ward councillor for this area, is involved in various social programmes throughout the city. She also developed Project Clean-Up, with which I have admittedly disagreed. However, in anticipation of the logical consequences of the cameras and PSC guards in the Bult area, she had the foresight to work with car guards who were about to lose their livelihood. As the cameras became fully operational the PSC and the CID started to encourage the public not to support any car guards in this part of the city. The councillor helped some to get documentation such as South African Identity Cards in order and lobbied many of the service providers, who took over services in the CID to appoint them. This was not strictly required. It speaks to a type of compassion that must be lauded and acknowledged. There should be more people doing this type of productive work, which can help to gradually articulate greater solidarity and collaboration into the social order. Especially when the state is remiss in many of its most vital functions, this type of initiative could reduce some instances of criminality by securing livelihoods.

Unfortunately, not everyone making a livelihood in the area that became the CID could be absorbed by the aforementioned service providers. Some have criminal records, which very likely played a role. While not hiring someone with a criminal record might be seen as reasonable, it can also be questioned. Someone needs to hire these unemployed residents, otherwise the odds of recidivism will likely increase. Failing to hire former convicts and taking away informal social security systems such as car guarding or introducing gatekeeping mechanisms, such as Project Clean-up did is questionable. As is mentioned earlier, the same person who instituted this gate-keeping mechanism in the Bult area has also demonstrated extraordinary compassion. Consequently, there are two points I would like to make and reiterate. This discussion is not about 'calling out' any person or institution. Such scholarship too often misses the mark in terms of ethics and even more importantly in opening up a space for meaningful deliberation and even respectful debate. 'Fierce' critique often alienates potential allies who are indispensable in creating a better social order. Secondly, neither people, nor institutions are homogenous entities. They are more complex. Rearticulating, slowly, but consistently the politics of everyday life can shape contexts in which different facets of people and institutions come to light. Unfortunately, based on a reading of existing literature on CIDs and enclavism, the cameras and the institutional structure of the CID has created a context in which some people can, informally, be banished by rendering and labelling them superfluous to the needs of the area. This might not have been the intent. The consequences may however be quite cruel for some, as the CID as a security infrastructure is (in)advertently serving the politics of closing the metaphorical frontier through biopolitical abandonment. Many of these car guards have had to deal with a lot of hardship. I have seen how the public treat them and through informal discussions learned about the pittance they actually made working extremely long hours. In the middle of winter in 2018 one car guard (Interview, June 2018) explained to me that he would spend three consecutive days and nights in the Bult area guarding cars until he had made enough money to make it worthwhile for him to return to his wife and children in Ikageng. In one of many incidents two young white women pulled up in a Sports Utility Vehicle (SUV) in the parking lot next to my car. A car guard asked the driver whether he could look after the vehicle while she and her friend were away. The driver responded by saying that she did not even have five Rand to buy herself an ice cream. She certainly did not have money to give to him. This example illustrates the ignorance by some of the realities facing many South Africans and the complete lack of compassion on the part of a young white middleclass person. I have seen many examples like this and I have through interactions with many other white and middleclass people, heard many frustrating arguments around how unemployment and homelessness are choices. This is exactly why I have cited the previous example of the initiatives by the local councillor as inspiration for the hegemonic rearticulation that Potchefstroom and arguably the broader South African social order is in dire need of. Besides car guards there is another common South African figure whose precarious livelihood has also been dealt a blow by the form that the CID has taken. Specifically, the large-scale recycling project when read in conjunction with my field notes from hours of direct observation in 2018 is not entirely positive.

The figure of the *bergie* is a familiar one in South Africa. There really is no meaningful translation for this concept. These are typically homeless people who live out of a supermarket shopping cart, which they push wherever they go. In it they keep all of their possessions and importantly, the recyclable materials that they gather out of garbage bins and refuse bags left in the street for pickup by residents and businesses. By dealing directly with a recycling company, structures such as the CID and the Oewersig residents' association essentially cut out this middle man. I have seen on various occasions people collecting paper and tin cans out of rubbish bins in the Bult area. The figure of the bergie, might be linked to the *sakskeerder* (bag shredders) or bin scratchers discourse observed in Potchefstroom online forums and in the literature on CIDs elsewhere in the country. This is not unique to Potchefstroom and other studies have revealed similar frustration amongst the white middle class. As Schuermans (2016:188) notes the Vredekloof Safety Council managing a residential CID in Cape Town, takes steps to discourage or, based on the aforementioned literature review, apply friction to homeless people, beggars and 'bin scratchers' from entering the neighbourhood. The Cachet Park CID frequently promotes the recycling company they employ on its Facebook page. By doing this and because of the general state of the public discourse and similar actions by neighbourhoods such as Oewersig, opportunities for paper, plastic and metal collection out of bins and refuse bags drying up significantly. Such a situation might counter-intuitively increase illegal recycling practices through recycling of stolen goods. A lack of available recyclable materials might also lead to other forms of theft. Based on the analysis presented in this section, the CID might want to consider all that is at stake. There is the security of students, workers and residents within the CID in relation to crime. This is a legitimate and important concern that arguably needed intervention. In addition, there is also the livelihood, food and nutritional security of some of the most vulnerable members of a shared social order. Due consideration should be given to spaces for interventions where these two interests are not mutually exclusive. Furthermore, is the reasoning that possibly frames these interests as mutually exclusive always sound or is there space for more nuanced intervention?

Narrow interpretations of nationalism

The CID as a physical and as an ontological bordered space may currently be interpreted as serving quite narrow notions of nationalism. Some might cite good reasons why this space serves Afrikaner interests in particular. They may cite the annual *Aardklop* Afrikaans arts festival in the park, which precedes the CID by decades.[2] However, there are additional forms of place-making that might be viewed as problematic. These include other events in the park, such as *boeremarkte* (farmer markets) (Cachet Park CID Facebook page 25 October 2019 and 15 October 2019). Here the English language idea of a farmer's market can be misleading, given the cultural associations historically attached to such events. *Boeremarkte* might reasonably be interpreted as events for *boere*, as mentioned in Chapter 2, a word that may also connote Afrikaner. The live entertainment advertised for these events, especially, support such a reading. Artists performing are very much of the taste one would associate with an Afrikaner identity. Finally, the CID's Facebook page has on various occasions posted materials associated with Afrikaner nationalism. These include references to the former *Zuid-Afrikaansche Republiek* (South African Republic), which essentially became the Transvaal Province of the Union of Africa after the South African War of 1899–1902 (cf. CID Facebook page 10 January 2020 and 7 January 2020). By posting such content the CID is not doing itself any favours in endearing it to the broader community. None of what I am saying should be taken as a stab against white Afrikaans cultural practices. I too belong to this group. The question is just whether these cultural practices and forms of identification should be reflected in the CID, be it deliberate or inadvertent. The problem is compounded by the obvious socio-economic differences between the CID and surrounding areas on the one hand, and other parts of the city on the other. There is a problematic exclusion of some in the name of security, alongside exclusion though culture and the socio-economic status.

Given the apparent emergence of a somewhat narrow nationalism in the unfolding articulation of the CID, it might be argued that two senses of order are at play in the politics of security in this part of the city. Order as a common term in Criminology can be associated with control over the movements of potential criminals and instilling a patterned set of practices that discourage crime. This is the first notion of order mobilised in the CID. It is often associated with theories of and approaches to crime such as Broken Windows, Zero Tolerance and target hardening through security infrastructures (cf. Wolf and Intravia, 2019; Gau and Pratt, 2010; Harcourt, 1998). While the extensive use of infrastructure is clear and it is believed to be effective, there have thus far been significant hints at the former two approaches. The CID has on some occasions stated that it will not tolerate infringements of traffic violations, such as not stopping properly at intersections (Cachet Park CID Facebook page 10 April 2020 and 21 January 2020).

The second notion of order deals with ontological security. Ontological security, to reiterate, refers to a sense of certainty and continuity about one's position within the social order. Here the public feed into the discourse significantly through complaints over out of place 'others'. Complaints over black minibus-taxi drivers who pick up workers in the area in the parking lot next to the park is one example. Such complaints are bundled with traffic offences, although the subtext might often be interpreted as divisive. These taxis are driven by and transport black South Africans. The same holds true for the discourse on car guards and 'bag shredders'. The limited range of events within the park, which might be construed as a somewhat regressive form of nationalism is another example. Both notions of order are problematic to varying degrees and for different reasons, depending on how they are acted upon. There is insufficient space to comprehensively deal with such reasons here. What the literature discussed previously suggests is that the politics of security becomes extremely problematic when the second notion of order (ontological security) informs the first (managing the movements of potential criminals). There is some evidence, be it preliminary, of such articulations of security into the hegemonic practices of the CID. Consequently, this politics of security needs some rearticulation. The following section offers a preliminary, and admittedly still somewhat general, set of recommendations.

A revised politics

Given the social order that I have argued we should be striving for, an amended politics of security in Potchefstroom should include certain elements. The overarching politics is premised on a logic of inclusivity, cohabitation and the co-construction of the social order across hitherto reiterated lines of division. This ambitious notion, as Laclau explains in terms of 'revolution', is an incremental, iterative process of continuous rearticulation. In the shorter term, realising that a meaningful future is only possible through collaboration between citizens is good a start. Moreover, there should be an understanding that enclavism and the forms of frontier governance, such as that described in this chapter and the previous one, are not ideal. Frontier governance, as spatial technique aimed at dealing with the contradictions of the political economic technique of closing the frontier through biopolitical abandonment, precludes interaction. Instead the public at large should demonstrate solidarity for diverse contemporary struggles by seeking out members of 'opposition groups' or existing public and private organisations with whom they believe productive discussion is more likely. Once initiated this might hopefully breed more productive interaction. If we frame crime as a societal concern, as opposed to merely linked to the individual characteristics of a perpetrator, then more obvious options might open up for collaboration.

Some more specific suggestions regarding the CID might include the following. To begin with the CID might seek to develop a social responsibility programme that is not designed around frontier governance. Instead there should be activities within the CID that are clearly meant to engaged people otherwise abandoned. The purpose of security infrastructures should be security and not the exclusion of certain categories of people. The CID has no legal right to banish car guards or other people looking to eke out livelihoods in the city. The discourses of 'criminal car guards' and 'sakskeerders' should be abandoned and openly challenged from within the CID. Instead, those who do engage in criminality, be they car guards, shop owners or anyone else should be policed. The assumption of car guards and similar categories as a criminal or unwanted other, however, has no future in a substantively democratic order, characterised by shared progressive praxes and equitable formal and substantive opportunities.[3] Instead the rapid response to crime should be the reason for crime reduction in the area. The CID's management should reconsider what may be construed as culturally exclusionary events held in the park. As it stands, the park and much of the CID could reasonably be seen as a narrowly conceived cultural space. This is not ideal, given the aforementioned objectives. Some creative thought is needed on more inclusive, though of course still safe, events in the park, which will likely draw diverse crowds. I am not suggesting that the extremely lucrative Aardklop festival be removed. That would be unwise for various reasons, including a loss of important revenue for the city and potentially alienating the Afrikaans people from the more inclusive politics we should striving for. Festivals such as Aardklop have a legitimate place in South Africa. But, which events might the CID add to the park's current roster that will serve national-building more broadly?

Conclusion

The notion of a CID in Bult area is not irrational, especially in the contemporary discursive milieu. We may, however, ask some questions about the discursive milieu and its potential for a truly inclusive democratic order. Many of the interventions implemented through the CID in effect (in)advertently white washes the harsh realities of life in the JB Marks Municipality. These interventions often banish those deemed to be threats to white and middleclass ontological security, along with potential threats to crime, from this part of the city. Conflating security from crime and ontological security more broadly, in turn, poses significant threats to responsible, conscientious and reflective citizenship. Stating that this conflation of two different notions of order is especially problematic does not detract from the fact that overreliance on security infrastructures to reduce crime is not, in itself, problematic.

I have made some suggestions for the productive rearticulation of the social, given the reality of the Cachet Park CID. The CID as a significant security node within the city is in a position to influence how it is policed. It should ideally take a leading role, along with other actors in positions of relative influence to effect hegemonic rearticulation that aids inclusion and collaboration across reiterated fault lines. Disentangling the loaded and floating signifier of security to break with what Lemanski (2006) calls 'fear of crime plus' will not necessarily be at the expense of the safety of students, residents and workers. When carefully thought out and seen as subject to continuous deliberation and recalibration, such a break can help to reduce crime, while building good faith relationships that might be of use in addressing other shared concerns. People deserve to feel safe. At the same time creative ways should be sought to make safe spaces inclusive spaces.

Notes

1   It might be noted that neither the UP nor UCT's campuses are fenced off from the rest of their environments in the way the NWU's campus is.
2   We may loosely translate Aardklop as the Earth is moving to a beat. The word suggests an exciting vibrancy one might associate with an arts festival in this instance.
3   These formal and substantive opportunities for democracy at a very basic level may include upholding constitutional rights. More substantively, these spoils of democracy may include equitable opportunities for the co-creation of a city that better serves the needs of a broader population.

References

Berg, J. 2013. Governing security in public spaces: Improvement districts in South Africa, in R.K. Lippert and K. Walby (eds.) *Policing cities: Urban securitization and regulation in a 21st century world.* New York: Routledge.

Besteman, C. 2008. *Transforming Cape town.* Berkeley: University of California Press.

Broadbridge, H.T. 2001. Negotiating post-apartheid boundaries and identities: An anthropological study of the creation of a Cape Town suburb. Unpublished PhD thesis. Stellenbosch University. Stellenbosch.

Clarno, A. 2013. Rescaling white space in post-apartheid Johannesburg. *Antipode: A Radical Journal of Geography*, 45(5), pp. 1190–1212.

Didier, S., Peyroux, E. and Morange, M. 2012. The spreading of the city improvement district model in Johannesburg and Cape Town: Urban Regeneration and the neoliberal agenda in South Africa. *International Journal of Urban and Regional Research*, 36(5), pp. 915–935.

Gau, J.M. and Pratt, T.C. 2010. Revisiting Broken Windows Theory: Examining the sources of the discriminant validity of perceived disorder and crime. *Journal of Criminal Justice*, 38, pp. 758–766.

Gerber, J. 2018. SA has 'deficit' of 62 000 police officers – Sitole. *News24*. 11 September. Online: www.news24.com/news24/southafrica/news/sa-has-deficit-of-62-000-police-officers-sitole-20180911. Date of access: 15 July 2020.

Hannigan, J.A. 1995. The postmodern city: A new urbanization? *Current Sociology*, 43(1), pp. 152–214.

Harcourt, B.E. 1998. Reflecting on the subject: A critique of the social influence conception of deterrence, the Broken Windows Theory, and order-maintenance policing New York style. *Michigan Law Review*, 97(291), pp. 291–389.

Hart, G. 2013. *Rethinking the South African crisis: Nationalism, populism, hegemony*. Athens: University of Georgia Press.

Kempa, M. and Singh, A.M. 2008. Private security, political economy and the policing of race. *Theoretical Criminology*, 12(3), pp. 333–354.

Lawson, V. and Elwood, S. 2014. Encountering poverty: Space, class, and poverty politics. *Antipode: A Radical Journal of Geography*, 46(1), pp. 209–228.

Lemanski, C. 2006. Residential responses to fear (of crime plus) in two Cape Town suburbs: Implications for the post-apartheid city. *Journal of International Development*, 18(6), pp. 787–802.

Lippert, R.K. 2014. Neo-liberalism, police, and the governance of little urban things. *Foucault Studies*, 18, pp. 49–65.

McKenzie, E. 2011. *Beyond privatopia: Rethinking residential private government*. Washington, DC: Urban Institute Press.

Paasche, T.F., Yarwood, R. and Sidaway, J.D. 2014. Territorial tactics: The socio-spatial significance of private policing strategies in Cape Town. *Urban Studies*, 51(8), pp. 1559–1575.

Rink, B.M. and Gamedze, A.S. 2016. Mobility and the city improvement district: Frictions in the human-capital mobile assemblage. *Mobilities*, 11(5), pp. 643–661.

Saayman, M. 2019. Student vertel van mesaanval. *Potchefstroom Herald*. 24 January.

Saayman, M. 2018. Student verkrag na gedokterde drinkie. *Potchefstroom Herald*. 26 July.

Samara, T.R. 2010. Order and security in the city: Producing race and policing neoliberal spaces in South Africa. *Ethnic and Racial Studies*, 33(4), pp. 637–655.

Sandhu, A. and Haggerty, K.D. 2016. Private eyes: Private policing and surveillance, in R. Abrahamsen and A. Leander (eds.) *The Routledge handbook of rivate security studies*. London and New York: Routledge, pp. 100–108.

Schuermans, N. 2016. Enclave urbanism as telescopic urbanism? Encounters of middle class whites in Cape Town. *Cities*, 59, pp. 183–192.

Valentine, G. 2008. Living with difference: Reflections on geographies of encounter. *Progress in Human Geography*, 32(3), pp. 323–337.

Van Riet, G. 2020. Intermediating between conflict and security: Private security companies as infrastructures of security in post-apartheid South Africa. *Politikon: The South African Journal of Political Studies*, 47(1), pp. 81–98.

Wetdewich, S. 2018. Rooftogte of the Bult word meer gevaarlik. *Potchefstroom Herald*. 31 May.

Wilson, H.F. 2011. Passing propinquities in the multicultural city: The everyday encounters of bus passengering. *Environment and Planning*, 43(3), pp. 634–649.

Wissink, B. 2013. Enclave urbanism in Mumbai: An Actor-Network-Theory analysis of urban (dis)connection. *Geoforum*, 47, pp. 1–11.

Wolf, K.T. and Intravia, J. 2019. Broken windows/Zero-tolerance policing. *The Wiley Blackwell Encyclopedia of Urban and Regional Studies*, pp. 1–6.

Interview

Car guard. 2018. Interviewed by Gideon van Riet. 16 June. Potchefstroom.
Car guard. 2018. Interviewed by Gideon van Riet. 5 May. Potchefstroom.
Head of Campus Protection Service. 2020. Interviewed by Gideon van Riet. 24 January. Potchefstroom.
Shop owner. 2018. Interviewed by Gideon van Riet. 16 June 2018.

E-mail

PSC manager to Gideon van Riet. Subject: 'Onderhoud'. 2 March 2021.

Concluding analysis
Crime, radical and plural democracy and strategies of construction

Introduction

The argument presented in this book still requires some concluding analysis based on the various threads drawn from previous chapters. Therefore, this concluding chapter is structured into three sections. The first section summarises the book's main arguments and conclusions thus far. The second section provides an analysis that builds and concludes on the preceding summary. This analysis concludes by calling for strategies of opposition, that is, radical and plural democracy, augmented by strategies of construction (Laclau and Mouffe, 2014[1985]:173). In both instances the suggested praxis will stem from collaboration between historically conflicting groups. These are required to invoke a more comprehensive response to crime in light of limited efficacy, for various reasons, within the state. South Africa must dispense with the political economic logic of closing the frontier through biopolitical abandonment laager and the spatial logic of frontier governance and its pivotal edifice of the laager, to render a new hegemony. Finally, the chapter considers the prospects for generalisation from this case study of a small South African city. As such, the chapter also concludes on the implications of this study for the fields of Security Studies, private security studies and the study of infrastructure, by social scientists.

Preliminary summary

Part I of this book established a macro-level interpretation of South African social orders. As such, it sought to contextualise security infrastructures in Potchefstroom. Chapter 1 elaborated the metaphor of a re-opened frontier and the logics of closure through abandonment and frontier governance in South Africa. It conceived of the ideal-types of the political and economic elites in contemporary South Africa. The chapter also offered an, admittedly partial though significant, interpretation of crime in contemporary South Africa as largely related to marginality, which is often expressed through variables such as multidimensional poverty and importantly, inequality and the expectations and associated senses of injustice that have accompanied the post-apartheid social order. Therefore, frontier governance and crime reduction premised on self-contradictory attempts

DOI: 10.4324/9781003028185-14

at closing the frontier through biopolitical abandonment are self-defeating. Secondly, Part I explored border-making in contemporary Potchefstroom. Chapter 2 introduced the laager as a concept for security analysis. It was conceived of as the sum total of physical and ontological structures that facilitate border-making. In Laclauian terms, the laager is a pivotal collective infrastructure of various nodal points through which actors articulate difference. Chapter 3 dealt with territorial stigmatisation as a means of border-making. By stigmatising Ikageng and some parts of Potchefstroom as sources of insecurity, the various insides of the laager have been justified as needing to be insulated from an outside largely constructed based on logics of equivalence. Part I, setting the scene of a metaphorical re-opened frontier and a physically and otherwise fortified laager therefore, also briefly touched on the problem of divisive discourses that are attached to the legitimate concern of crime. These discourses often explicitly or implicitly manifest as 'us' and 'them', based on ideas such as race, nationality, socio-economic status and livelihood strategies.

Part II further developed an understating of crime discourse as entangled with additional problematic ideas and practices. This part of the book focussed on the echo-chamber effect of crime discourse. Chapter 4 dealt with the CPF. Chapter 5 analysed the contents of the *Potchefstroom Herald* and Chapter 6 analysed the contents of PSCs' social media platforms. Each of these infrastructures overlap in terms of the content and underlying messages that flow through them. In all three cases the institution in question does not always assess the potentially hazardous nature of content. However, Facebook group administrators and newspaper editors often allow problematic messages and innuendos by the public to be published. These ancillary discourses may productively be conceived of as floating signifiers that may be challenged in order to limit the social division that derives from the problem of crime. Due care is required to combat crime and alleviate fears. An important argument of this book has been that these two imperatives must as far as possible be approached as mutually reinforcing and not mutually exclusive.

Part III built on the largely ideational aspects described in Part II by explaining how these are manifested through routine practices. Part III also fleshed out the overarching notion of the laager by emphasising its physical manifestations at lower scales. In particular, Chapter 7 dealt with PSCs and their employees who patrol middleclass neighbourhoods. Chapter 8 investigated the Oewersig Residents' Association, as a pilot project in the city, where residents took over certain municipal functions. These residents augmented security provision in their neighbourhood through CCTV cameras in the streets monitored live by a PSC. Other neighbourhoods are following suit. Finally, Chapter 9 investigated the newly formed Cachet Park CID, which employs many of the same practices and technologies as the Oewersig Residents' Association. This partnership between the North-West University, the city council, business owners and residents, contracted out landscaping, refuse removal and recycling as well as security provision. The same PSC employed by the Oewersig residents, monitors the CID through foot patrols, a dedicated armed response vehicle, a mobile incident reporting centre and well over 100 CCTV cameras across a limited number of city blocks. The physical and practical manifestations of the laager are intertwined with the ideational and ontological manifestations described in Part II. As with previous chapters, Part III further developed the Laclauian notion of the non-uniformity of identities, including those involved in border-making. These subjectivities include armed responders and white and middleclass municipal councillors. Some of the most significant adverse consequences of security practices identified in Part III, pertain to limits to freedom of movement, the elimination of certain sites where livelihood strategies have been developed by some of the most vulnerable members of society, the active development and maintenance of exclusive spaces of representation and the likely displacement of crime. The latter issue reinforces and deepens the trend of exclusive security communities (Gheciu, 2018:14;28), who believe they deserve special security privileges, because they can pay. In addition, any notion that such security infrastructures reduce crime is premised on the idea that crime is a property of the individual, who can be prevented from engaging in criminality, and not a problem of a social order that produces excessive marginality through biopolitical abandonment. In regards to many of these adverse consequences the economic elite are the most active advocates, but the political elite are at least negligent and therefore also complicit.

Analysis: radical and plural democracy and strategies of construction

The notion of hegemony employed in this book is different from more common usages of the term, typically based on the work of Gramsci (1971 edition). Unlike Gramsci a Laclauian perspective on hegemony rejects the idea of a hegemonic centre. Sovereignty is not so much moved away from the state as it is hegemonic formations within the nation-state that have and continue to be re-articulated. A Laclauian perspective on hegemony also rejects the idea of a neat antagonistic dichotomy within society. Radical contingency means that perceived dichotomies are the result of logics of equivalence. Dichotomies are therefore produced. Equivalences can be problematic, but they can also be productive for a pragmatic politics in the service radical and plural democracy, which I have, following Laclau and Mouffe, defined as diverse and fluid means whereby constituencies collaborate to hold the state accountable.

There is a productive tension between logics of equivalence and logics of the difference. The first collates various individuals and groups into one category for the sake of defining an inside and an outside. The second recognises the diversity within a social order and its constitutive sutured categories. The politics of radical democracy entails calling out unproductive and oppressive equivalences. At the same time equivalences may be deployed for the sake of resistance and continuous articulation, that stands in place of a sudden radical Jacobin revolution (Laclau and Mouffe, 2014[1985]:61). Critical imperatives will change over time. Instituting the logic of a radical and plural democracy, which embraces the productive tension between equivalence and difference could better serve changing interests, whilst, if carefully and consistently driven, sustaining a continuous progressive politics. And, it seems difficult to argue that such a politics is not required in one of the most enduringly unequal material and otherwise polities in the world.

It should be recognised that hegemony is never 'overcome'. It is simply continuously re-articulated. Hence, hegemony is neither purely good or bad, but some versions can be more useful than others. Laclau and Mouffe (2014[1985]:161) argue that states cannot be a panacea for intervention. Moreover, equivalences are not the only or primary political spaces and categories. Also, the aforementioned plural democracy, which might be viewed as *analogous*, though not identical, to philosophical anarchism (not to be confused with 'chaos'), implies multiple praxes by diverse means on different fronts. Many of these political projects should be about holding states accountable. Furthermore, South Africa as a state that looks well after all who live in it has never existed, a point I will return to shortly.

Throughout I have tried to be sympathetic to diverse interest groups. By applying a largely Laclauian framework, I have tried to avoid what Giddens (in Rantanen, 2005:73) calls 'infantile leftism' or perhaps we may rephrase this as 'infantile constructivism'. Crime is real and people do have good reasons to be afraid. Moreover, crime poses various developmental challenges. It discourages investment (Cheteni et al., 2018:1), which in turn might inhibit job creation, to name only one example. We cannot expect people to give up their security infrastructures overnight, when they feel unsafe. In this book I have tried to understand what this fear does; how it interacts with other discourses and eventually impacts upon the post-apartheid South African project. The laager and the city as structures of contemporary hegemony are dynamic. Security infrastructures partially shape this dynamic. What is consistent, however, is the absence of any meaningful right to the city for the majority (cf. Lefebvre, 1996[1968]; Harvey, 2012). What I mean by the right to the city, is the *substantive* freedom and ability to live in the city and 'live in it well' (see Purcell, 2013). As such, this right includes the likes of legitimate livelihood opportunities and freedom of movement. Unfortunately claims to victimhood from positions of relative privilege legitimise the dehumanisation of others and hampers social cohesion. A view of crime, entangled with problematic, though very different compared to the 1980s', notions of 'onslaught', facilitates retreat and insularity. It also limits constructive engagement with authorities and between residents of different strata in addressing the crime problem.

Throughout this book I have emphasised that South Africa requires a comprehensive or holistic approach to crime reduction. Such an approach includes a macroeconomic component, microeconomic components, remedial spaces outside of the criminal justice system and diverse partnerships between the state and private actors, but also amongst diverse private actors. The current emphasis on target hardening, environmental design and security infrastructures, such as alarms, fences, and cameras cannot possibly deal with the problem of crime and the social divisions that accompany it. Much more needs to be done to arrest multidimensional and complex matters such as unemployment, homelessness and inequality. Some initiatives of this order are possible through each municipality's five-year

integrated development planning cycles, although these have historically struggled to gain traction in many contexts.[1] Based on the discussion thus far some
obvious and less obvious conclusions can be drawn.

It seems quite clear that South Africa has to revisit the approach of the now
defunct National Crime Preventions Strategy (NCPS) of 1996. Authorities should
however in the process, heed the lessons that have been learnt from Development
Studies, regarding such notions as 'mainstreaming' and inter-sectoral collaboration (see van Riet, 2017:22–23). Authorities and key actors, such as the SAPS
should embrace the principles the NCPS was based on. They should however be
less ambitious in terms of inter-sectoral collaboration. Instead, the priority should
be to build up the various departments that may be of benefit to crime prevention.
Success in this regard, will likely have some bearing on crime. Another obvious shorter term intervention might be to heed the suggestions of others, such as
the return of special units within the police that deals with widespread problems
such as drugs (Cheteni et al., 2018:13). Many of these units were dismantled a
number of years ago. As they were previously somewhat independent of local
stations, their reinstatement might reduce the potential for corruption and importantly perceptions of corruption. In the shorter term, existing structures such as
the SAPS, CPFs and responsible private security should be strengthened as far as
possible, given the limitations expressed for example in Chapter 4. In the longer
term however, the structure and rules of the broader social order needs continuous
re-articulation.

The problematic of crime in Potchefstroom, much like other parts of South
Africa, is both a consequence of and echoes the deeper problematics of inequality, poverty and social cleavages in the overarching social order. Crime is not a
problem that can be ascribed to individual criminals exclusively. It is a problem
for and of the broader social structure. Meaningful primary responses can therefore not be, border-making, target hardening and overloading the criminal justice
system. Often these types of interventions fuel the larger problem. It entrenches
a mode of thought whereby security can be bought through a monthly subscription, when the reality is that it is the social and material distance between South
Africans that primarily drives crime. This problematic is compounded by the fact
that South Africa has never, neither during nor after apartheid, had a government
that truly expressed, through policy implementation, cared for the plight of the
average citizen. There are not many signs that this will change anytime soon.
A more likely source of promise lies in the people themselves, but only if they are
willing to act by reaching across reiterated cleavages, to affect change by themselves, and to form collective forces that counter-balance and place pressure on
indifferent governance. These collective forces might be a plurality of alliances
on diverse issues, for our purposes those that directly or indirectly affect crime.
What is important, is the recognition of the openness of meaning-making and
collaboration across these fault lines. Hegemony has thus far been tied to reiterations of conservatism and not the required re-articulations towards progressive
ends. Progressive re-articulation will be incremental, but it should be unrelenting,
as residents develop new modes of thought through continued socialisation. The

need for coalitions across the likes of race and class is essential, to demystify the equivalences to which they are attached. The problem of crime is a collective one, which implies much need for collaboration. As such, crime is an excellent example of a problem fuelled by the very equivalences used to respond to it.

Despite the problems of municipal service delivery and continuous protests in many contexts, corruption and often lacklustre macroeconomic strategy and execution, the ANC has remained in power. Desai (2018:508) makes the following observation.

> What does it say about the character of the ANC that it could keep Zuma for so long? In some respects, the answer partly flows from the hold that a leader playing the victim, playing up race and insinuating imperialist plots has over the national psyche. Because of our history, these tropes have an enduring material basis.

This observation suggests two things. Firstly, there are major problems within the ANC. The factional politics of individual survival within the party is stifling in all spheres of government form the local to the national. At the local level an apparent fusion between party, state and large parts of civil society complicates the argument I am putting forward. Cadre deployment, factionalism within the ANC and protest politics are often linked by a politics of patronage. Local protests, contrary to media depictions are not always about holding the state accountable for services. Often, these protests are a manifestation of factionalism and the politics of loot in a failing economy. This patronage extends far beyond large tenders. They may also involve access to low paying, even precarious, job opportunities (cf. Von Holdt, 2013; Alexander, 2010). In this regard, ANC politics and political opportunism has in many cases penetrated both invited (participatory forums) and invented (protests) democratic spaces (cf. Sinwell, 2011:63). Thus, collaboration between diverse segments of the social order will have to carefully approach this politics of basic survival for some linked to the politics of careerism within the ANC. Still, this complex link between state, party and civil society too is dynamic and thus not impenetrable. The fact that the ANC appear to have lost votes in recent years, be it to apathy and not opposition parties in many protest prone areas is a positive sign (Ruciman, 2017:1). The fact that authors such as Hart (2019) links unrest to inequalities tied to neoliberalism supports the praxes proposed in this book and elaborated in the following.

Secondly, South Africa has a particularly divisive recent history, which has been reiterated, for example in the form of incessant racism. Consequently, the economic elite cannot expect the majority of citizens to vote out the ANC in the broader JB Marks Municipality or at national level. At a minimum, it is unlikely that these citizens will vote into power their preferred party, typically the Democratic Alliance (DA). The DA is the only major party not significantly tainted by recent revelations of corruption. Yet, that party in its post-apartheid guise is more inclined to (neo)classical liberalism and associated economic policies. The DA also holds connotations of being a white party, serving white interests. There was a moment of promise of a remodelled and more progressive DA under the former leader Mmusi Maimane. But he was not allowed to continue on this path because of a loss of voters during the 2019 national elections, mostly to the right. The 2019 elections and its aftermath for the DA might very well be seen as yet another in a long history of missed opportunities for the country. Other similar missed opportunities include the abandonment of ISI when it was still accepted globally in favour of continued apartheid, the outcomes of the negotiated settlement of the 1990s and the fraught rhetoric of an emerging but never materialising 'developmental state'. It is a real pity that this loss of voters to the right was viewed by the party establishment in negative terms. It could have marked the beginning of something far more progressive, a party relatively untainted by corruption, moving towards alignment with the interests of a broader segment of South Africans. So, if party politics does not offer a viable means of directly or indirectly intervening into the problem of crime, what does?

Perhaps crime reduction may be achieved regardless of who is in power, provided leaders are under pressure from multiple groups engaging in radical and plural democratic politics. These forms of politics would include a range of South Africans who exhibit a degree of collaboration for common interests. While a complete end of stratification is inconceivable today, it should be quite clear that there are limits to how much stratification a social order can tolerate, before it capitulates. It is for this very reason that 'South Africa' has always been a precarious notion. For all of the reasons cited earlier, there seems to be very little long-term options for those in positions of relative privilege outside of seeking partnerships with people many of them despise for voting for the ANC. This partnership is no longer the mere 'luxury' of an ethical imperative, which it has always been. The South African crisis as described by Hart (2013) and articulated somewhat differently by myself (Van Riet, 2020), arguably now requires more forceful resistance to indifferent governance, where the power of numbers and the power of economic and political clout align. This perspective is not completely unique. Finn (2020) makes a similar argument in the context of South Africa, which he uses to critique Chatterjee's (2004) distinction between political and civil society. Here civil society refers to the segment of the population more typically active in invited democratic spaces and spaces between state and the household as per the definitions of these concepts in the Western Political Science canon. Political society refers to those subjected to neoliberal governance, in a sense the abandoned majority described in Chapter 1. These residents live of the margins of 'society' and struggle to survive by purely legal means. Finn advocates for a new popular sovereignty in South Africa. Popular sovereignty is a historically contingent and forward looking notion (Finn, 2020:5). Finn agrees with Mbembe (2015) that South Africa is a country characterised by persistent racial and economic silos and that these silos need not be permanent. He argues that Chatterjee's distinction between civil society and political society is underdeveloped. These seemingly distinct categories are part of a single dynamic. Moreover, reiterating such a distinction may reify a presumed gap between civil society and political society, thereby entrenching divisive politics (Finn, 2020:11–12). By employing the Chatterjean dichotomy a lack of democracy becomes something that cannot be overcome. As such, logics based on the distinction between civil and political society implies the absence of the hope the current study has been at pains to salvage from the politics of crime and security in Potchefstroom. Instead Finn argues for a continuum of political sovereignty, which following the logic of my analysis (re)introduces hope defined as the potential for greater collaboration across hitherto seemingly absolute lines, characterised by the equivalence of diverse struggles in more or less egalitarian terms. The link between civil society and consolidated democracy therefore lies at the level of socialisation. Realising some of this hope would imply steering contemporary politics away from its most divisive manifestations. A likely starting point would be those within elite groups more inclined to such partnerships. With time more people may be recruited and a critical mass, one day, may pivot the social towards more sustained open-ended progressive praxes and a new hegemony stemming from the productive tension between equivalence and difference. As I have demonstrated there is sufficient diversity within sutured categories such as those comprising or working for the economic elite to at least make a start. We now move to the main conclusion of this research, before the final section, which contemplates its implication for particular fields of study.

Playing on the productive tension between the logics of equivalence and difference might be a useful start as sources of opposition to crime trends and associated socio-economic concerns. New continuously reiterated and thus always partially sutured, hegemonic projects, however, need nodal points, by which it can be defined. Therefore, 'strategies of opposition' should be accompanied by 'strategies of construction' (Laclau and Mouffe, 2014[1985]:173). It is truly imperative that these strategies of positivity, that is, strategies aimed at shaping new social orders, are the result of continued partnership. The exact substance of the projects that constitute the social and the membership of collaborations may of course vary continuously. Eventually we may institutionalise a 'new common sense', 'where struggles are not wholly sutured competing interests' (Laclau and Mouffe, 2014[1985]:167). For this to happen, contemporary modes of physical, practice based and ideational border-making must see sustained movement towards its eventual exclusion from future hegemonic orders.

The contribution of this book and matters of generalisation

A few questions remain, pertaining to the relevance of this work to other contexts. The mode of analysis employed in this book is not typical within Security Studies or Criminology. I pointed out these differences in the book's introduction. By focussing on hegemonic articulations in the form of ideas and practices, directed through physical and metaphysical infrastructures, the book has demonstrated how security is rooted in its social and socio-physical context. This context is shaped at different scales, but the influence of higher scales on lower scales is inconsistent. Potchefstroom might to some extent be an archetype on account of the demographics within this particular police precinct. It might therefore seem as if the laager as a collective of infrastructures, historically a trope associated within

Afrikaner insularity and racism may have limited mobility. The laager has, however, been conceived of as a collective of physical and ontological structures that serve border-making in response to a fear of crime *and* relative position within the social order. Based on this definition, my response to questions of mobility would be that laagers could manifest in different contexts, be it with local inflections. In South Africa, Potchefstroom is not unique. I would imagine that as part of a smaller city, Potchefstroom might also be comparable to other, especially settler colonial contexts. Most studies on crime and security have been in large metropolitan areas. The literature cited throughout the book, is very clear on similar forms of insularity in large cities. The only questions, relate to the nature of the ancillary discourses attached to crime, its relationship to demographics and how these are borne out in the form of a laager. I would furthermore hypothesise that similar ancillary discourses have been internalised by some wealthier people of colour. In any event, it seems quite likely that the laager, as defined earlier, possibly has heuristic value elsewhere inside and outside of South Africa.

In terms of understanding crime and security in South Africa, the book has revealed the extreme difficulty of addressing crime. There is a tension between the economic power of elites, their fears and the rights of others. My argument has been that this tension is largely premised on a false set of logics of equivalence, in terms of who to fear and who might conveniently be kept out of sight for reasons such as the inconvenience of being confronted with abject poverty as a relational manifestation. In addition, the supposedly uniform subjectivities of those on the 'inside' are also problematic. Strategically deployed logics of difference might also help to address another set of false assumptions. These assumptions include the belief that crime is purely a property of the individual perpetrator and not the social order more generally, and that the status quo is tenable indefinitely, through biopolitical abandonment, border-making and more (in this case mostly private) policing. I have been critical of private policing while acknowledging the limits in state policing capacity. I have also suggested diverse collaborative praxes between public and private actors and between private actors, such as crime reduction interventions beyond policing.

As for the study of private security, I have attempted to broaden our perspective on these actors by situating them in relation to various other security infrastructures and the public. There has been a lot of excellent studies on private security from within International Relations, Political Studies, Anthropology, Sociology and Criminology that have helped us understand these increasingly important political actors. The book has in a sense 'zoomed out' in terms of scope, but used a specific geographically defined case study to illustrate the social embeddedness of private security. This social embeddedness is relational. I have repeatedly invoked the spectre of Ikageng in this book. It has been a necessary outside by which the inside of the laager has been defined. Such outsides should be centred in future studies, so as to understand them on their own terms.

The book has also demonstrated the at times tenuous link between the study of security and International Relations. However, this is a statement in need of significant qualification. Of course, ontologically, all contexts are part of a global system with its own secular processes. However, drawing links across scales should not be forced. This is not because 'globalisation' is not real. It is because there is merit in studying micro-contexts. Security provision and political strategy happens across scales. Making an argument situated in an international imaginary might sometimes come at the expense of revealing important local level power relations and particularities, which might in turn still be of use to researchers working in other contexts. Yet, Security Studies has largely, though definitely not exclusively, remained tied to the international, presumably based on its roots in International Relations.

Finally, the study of infrastructure by social scientists has grown immensely in recent years. The book, has further developed the body of research. It has contributed to extending the notion of infrastructure beyond its traditional narrow conception. By combining a broadened infrastructural imaginary with a post-structural theory of hegemony a few relatively novel analytical spaces opened up. These relate to a focus on both the ideational and physical nature of flows and the underlying infrastructural substrate. Within this politics of crime, theorising the spatial politics of crime may be both critical and in need of a novel politics or praxis of desegregation and crime reduction. Future research, might assess this politics further, by analysing the flows that help constitute it, whilst more fully considering the implications and limitations of the ontological as infrastructure.

Note

1   All municipalities have to compile a master developmental plan every five years, which outline the developmental activities to be undertaken within that period. The plans are call integrated development plans or IDPs for short.

References

Alexander, P. 2010. Rebellion of the poor: South Africa's service delivery protests – a preliminary analysis. *Review of African Political Economy*, 37(123), pp. 25–40.

Chatterjee, P. 2004. *The politics of the governed: Reflections on popular politics in most of the world*. New York: Columbia University Press.

Cheteni, P., Mah, G. and Yohane, Y.K. 2018. Drug-related crime and poverty in South Africa. *Cogent Economics & Finance*, 6(1), pp. 1–16.

Desai, A. 2018. The Zuma moment: Between tender-based capitalists and radical economic transformation. *Journal of Contemporary African Studies*, 36(4), pp. 499–513.

Finn, B.M. 2020. The popular sovereignty continuum: Civil and political society in contemporary South Africa. *Environment and Planning: Politics and Space*, Online First. pp. 1–16.

Gheciu, A. 2018. *Security entrepreneurs: Performing protection in post-cold war Europe*. London: Oxford University Press.

Gramsci, A. 1971 edition. *Selections from prison notebooks*. New York: International Publishers.

Hart, G. 2019. From authoritarian to left populism? Reframing debates. *South Atlantic Quarterly*, 118(2), pp. 307–323.

Hart, G. 2013. *Rethinking the South African crisis: Nationalism, populism, hegemony*. Athens: University of Georgia Press.

Harvey, D. 2012. *Rebel cities: From the right to the city to the urban revolution*. London: Verso.

Laclau, E. and Mouffe, C. 2014[1985]. *Hegemony and socialist strategy*. London: Verso.

Lefebvre, H. 1996[1968]. Right to the city, in E. Kofman and E. Lebas (eds.) *Writings on cities: Henri Lefebvre*. Translated by E. Kofman and E. Lebas. Oxford: Blackwell Publishing. pp. 61–181.

Mbembe, A. 2015. Achille and Mbembe on the state of South African political life. Africa is a country. Online: http://africasacountry.com/2015/09/achille-mbembe-on-the-state-of-southafrican-politics/ Date of access: 20 August 2020.

Purcell, M. 2013. To inhabit well: Counterhegemonic movements and the right to the city. *Urban Geography*, 34(4), pp. 560–574.

Rantanen, T. 2005. The 'G'- word: An interview with Anthony Giddens. *Global Media and Communication*, 1(1), pp. 63–77.

Ruciman, C. 2017. The 'ballot and the brick': Protest, voting and non-voting in post-apartheid South Africa. *Journal of Contemporary African Studies*, 34(4), pp. 419–436.

Sinwell, L. 2011. Is 'another world' really possible? Re-examining counter-hegemonic forces in Post-Apartheid South Africa. *Review of African Political Economy*, 38(127), pp. 61–76.

Van Riet, G. 2020. Intermediating between conflict and security: Private security companies as infrastructures of security in post-apartheid South Africa. *Politikon: The South African Journal of Political Studies*, 47(1), pp. 81–98.

Van Riet, G. 2017. *The institutionalisation of disaster risk reduction: South Africa and neoliberal governmentality*. London: Routledge.

Von Holdt, K. 2013. South Africa: The transition to violent democracy. *Review of African Political Economy*, 40(138), pp. 589–604.

Index

Note: Page numbers in *italics* indicate a figure on the corresponding page.

Printed in the United States
by Baker & Taylor Publisher Services

Printed in the United States
by Baker & Taylor Publisher Services